Chernobyl

The Incredible True Story Of The
World's Worst Nuclear Disaster

(History Of A Human Disaster Life Death Rebirth)

Thomas Housel

Published By **Simon Dough**

Thomas Housel

All Rights Reserved

Chernobyl: The Incredible True Story Of The World's Worst Nuclear Disaster (History Of A Human Disaster Life Death Rebirth)

ISBN 978-1-77485-594-2

No part of this guidebook shall be reproduced in any form without permission in writing from the publisher except in the case of brief quotations embodied in critical articles or reviews.

Legal & Disclaimer

The information contained in this ebook is not designed to replace or take the place of any form of medicine or professional medical advice. The information in this ebook has been provided for educational & entertainment purposes only.

The information contained in this book has been compiled from sources deemed reliable, and it is accurate to the best of the Author's knowledge; however, the Author cannot guarantee its accuracy and validity and cannot be held liable for any errors or omissions. Changes are periodically made to this book. You must consult your doctor or get professional medical advice before using any of the suggested remedies, techniques, or information in this book.

Upon using the information contained in this book, you agree to hold harmless the Author from and against any damages, costs, and expenses, including any legal fees potentially resulting from the application of any of the information provided by this guide. This disclaimer applies to any damages or injury caused by the use and application, whether directly or indirectly, of any advice or information presented, whether for breach of contract, tort, negligence, personal injury, criminal intent, or under any other cause of action.

You agree to accept all risks of using the information presented inside this book. You need to consult a professional medical practitioner in order to ensure you are both able and healthy enough to participate in this program.

Table of contents

Introduction _____ 1

Chapter 1: Fast And Accurate _____ 5

Chapter 2: Explosions _____ 9

Chapter 3: The Pressure Of The Invisible Rays _____ 19

Chapter 4: The Fire Continued To Howl And Then To Howl _____ 24

Chapter 5: Don't Cause Panic _____ 29

Chapter 6: Come On, Come On _____ 49

Chapter 7: Required Bags, Shovels, Sand And People _____ 57

Chapter 8: The Dozens Of People Who Died _____ 62

Chapter 9: The Ash Fell On Pripyat _____ 68

Chapter 10: The Life-Changing Event ___ 81

Chapter 11: The Meltdown _____ 113

Chapter 12: Research And Studies On Chernobyl _____ 149

Conclusion _____ 176

Introduction

"The risks projections suggest that, as of today, Chernobyl may have triggered around 1,000 cases of cancerous thyroid as well as 4000 other types of cancer in Europe that is around 0.01 percent of all reported cancers since the incident. Modelling suggests that by 2065 around one thousand cases of cancer of the thyroid as well as 255,000 cases of other types of cancer could be expected to result from radioactivity from Chernobyl however, hundreds of millions of cancer cases could result from other factors." The findings were published in an article that was published by the International Journal of Cancer in the year 2006

Uranium is most famous for its destructive power the Atom Bombs that brought the nuclear age to an end towards the end of World War II, but due to the efficacy that nuclear energy had, it was no surprise that nuclear power stations were built throughout the industrialized world in the second quarter of the 20th century. Although these power stations were not an option and allowed for modern, more efficient and cost-effective methods of energy that could be used to be used for domestic consumption The use of nuclear energy certainly frightened people who lived during in the Cold War and amidst ongoing nuclear explosions. In the end, the destruction caused to Hiroshima and Nagasaki was a clear indication of the damage nuclear power could cause as well as the health issues that people who were who were exposed to radiation revealed the terrible adverse effects that could be triggered when using nuclear weapons, or the inability to harness the technology correctly.

The first major incident of a nuclear station was in Three Mile Island in Pennsylvania in 1979. It took more than 15 years and nearly $1 billion to get it cleaned up following the disaster However, Three Mile Island paled in contrast to Chernobyl that continues to be the most famous nuclear accident in human history. The Chernobyl power plant is located in Ukraine Chernobyl was Chernobyl was a nuclear power plant in Ukraine. Chernobyl nuclear power plant in the midst of tests in the early hours of April 26th the 26th of April, 1986 when it was struck by several explosions within one of its reactors. These explosions ended up which killed more than 30 people at the plant , and spreading radioactive fallout across a vast part across Europe and the Soviet Union. While the Soviets attempted to conceal how catastrophic the incident at Chernobyl was however, it was impossible to conceal the magnitude of the destruction, especially when radioactive matter was infecting Western Europe as well. In all, the disaster resulted in approximately $18 billion in damage caused by the accident, forced everyone to leave the area near and continues to cause adverse health effects continuing to be felt in the region.

Like Three Mile Island before it, Chernobyl emphatically demonstrated the dangers of nuclear power stations, and brought about the introduction of new laws across the globe in efforts to make usage of nuclear energy less risky. Scientists and researchers continue to study the effects of radiation on the people who have been exposed to it. They are also working to develop estimations of how dangerous Chernobyl will end up.

The Chernobyl Disaster is the most devastating nuclear catastrophe in human history as well as the aftermath of the disaster. With photos and an index, you'll discover more about Chernobyl unlike any other time in a matter of minutes.

Chapter 1: Fast And Accurate

Modern day image of Chernobyl plant taken from Pripyat, Ukraine

Photo of the reactor hall inside Unit 1 in Chernobyl

"What was it that Akimov and Toptunov as the two people who operated the nuclear process felt at the moment that the rods controlling it got stuck and the first shocks of terror were heard in the main hall? It's difficult to answer as both men passed away in pain from radiation and did not record any details about this event. However, one can imagine the pain they were feeling. I'm familiar with the emotions experienced by operators at the moment of an accident. I've been there many times as a worker in the operations of nuclear power stations. Within the first second you are numb an avalanche lands on your chest. You get a chilling feeling of fear involuntary at being awestruck and initially not being able to decide the best course of action when recording instruments' arrows and indicating instruments start flying across the sky and your eyes are trying to track them all at the same time while the reason and cause of the situation unclear, and simultaneously (again involuntarily) you're thinking in the depths of your mind, on an additional level about the responsibility and the implications of what transpired. In the following moment, you feel an unimaginable mental clarity and calm thinking. This means that your actions to locate the cause of the incident are quick and exact ..." Grigoriy Medvedev the deputy chief engineer of Unit 1 at the Chernobyl plant, and the creator of The Truth about Chernobyl.

On the afternoon of 25 April 1986 Grigoriy Medvedev sat in an aircraft flying close to the site of the biggest nuclear catastrophe of all time that compelled him explain the events that occurred before and after. Later, he wrote "We were flying over Ukraine that was then covered in flowers in its gardens. About 7 or 8 hours passed, and a new age would start for the region, one of devastation and nuclear contamination. While I was watching, I looked at the ground from the window. Kharkov floated below in a blue fog. ... Kharkov floated by below in a blue haze. magnificent Pripyat River! Its water is dark brown and brown, possibly because it comes from the Polesian peat bogs. The flow is swift and strong as you swim. it tried to drag your body away."

The people who lived in the region were not able to imagine they would have a different outcome during the early hours of the 26th of April 1986. People who were not asleep in the city of industry Chernobyl were mostly working night shifts of the nearby nuclear power station and a lot of the people who came to the facility the night before were aware that there was going to be a test of the systems on the same day. It was a routine procedure that was not given the test much thought at the time Aleksandr Akimov, the unit head of the shift, decided to begin the test just after midnight. He was married to Lyubov was later quoted as saying, "My husband was a extremely amiable and social person. He was very easy to get along with people, but he was not one for the familiarity. A kind man who discovered happiness in his life. He was active in civic service. He was an active member of the Pripyat Gorkom. He loved his sons extremely greatly. He was considerate. He loved hunting and especially after he had gone to work and we bought a vehicle."

Akimov

Prior to the start of the test the test was over, there was a major problem after a sudden surge of power occurred in the fourth reactor. The power fell from 1500 MWt down to 30 MW, leading Akimov to propose that the experiment should be canceled. But, Anatoly Dyatlov, the deputy chief engineer who was in charge of the test, demanded that they continue. Based on Viktor Grigoryevich Smagin, the shift chief of Unit 4, "Dyatlov was difficult to work with, one who operated with delayed action. He would often tell his subordinates that I do not execute a punishment in one go. I contemplate the actions of a subordinate for at least a few days until there is no sense of resentment or anger in my mind, I take the decision. ....' To do so, he had gathered several physicists as managers from Far East, where he was chief of a laboratory for physics. Orlov as well as [Anatoliy AndreyevichSitnikov, Sitnikov, (both killed) were also from the region. Many others were also acquaintances and colleagues from the place the former worker had worked prior to his death. The norm in the Chernobyl plant prior to the blast was to beat up the workers on their shifts, and to give and offer incentives to working during the daytime (non-operating) employees of the shop. There were more incidents at the turbine area and less within the department of reactor. This is why there was a more sluggish attitude towards the reactor. The impression was that it was safe and reliable. ..."

Dyatlov

Chapter 2: Explosions

"Once the water lines in the lower part of the reactor that coolant was pumped to the core, were ripped away, the reactor was completely devoid of water. Unfortunately, as we found out afterward, the managers didn't know this or didn't want to believe it. This led to a whole sequence of incorrect actions that led to over-irradiation and deaths that could be prevented. Thus, the explosions ...."
Grigoriy Medvedev

After a few minutes of the initial surge and the power level remained at 200 MWt. All was well however, nobody was aware at the moment that xenon is leaking to the nucleus, poisoning it and stabilizing it. As Akimov started the test at about 11:30 a.m. The main water pumps began to fill with bubbles triggered by the water inlet which was beginning to boil. The power generated by the reactor grew to dangerous levels as water surrounding the core evaporated and the core's depth dropped from 25 feet of depth to approximately 8 feet deep. Medvedev later stated "It was repeatedly stated at press conferences that, prior to the explosion, the reactor was shut down, and that the rods were placed in the core. However, as we've previously stated, the efficacy for the security system in emergency situations was in all actual senses invalidated due to the blatant infractions to the operating rules. When pressing the AZ button was activated the rods controlling the reaction were, as previously mentioned, only went 2.5 meters deep into the core, instead of the expected 7 meters and they didn't smother the reaction in the first place, but rather helped to facilitate the prompt-neutron excursion. The system did not hold one press conference on this glaring error made by the engineers of the system, which ended up being the primary reason for the nuclear accident."

In the hope of avoiding the disaster, Akimov disconnected the clutches which held the control rods up in the air. He hoped that their weight would drop them further into the water However, they were stuck. After a while the control rods began to tense up and the reactor burst into flames in a violent way, hurling objects throughout the area, and sending dust out into the air around the building, as well as knocking out the power.

In just a few minutes the room was filled with men who had reported that the top of the reactor was collapsing on itself and there was an fire within the hall for turbines. Akimov was the first to report the fires and then contacted the electrical department asking for the power needed to manage the blaze, but the lines to the telephone went down. Then, Dyatlov did not believe the nuclear reactor had ruptured, so he demanded the reactor to be cooled by the help of a torrent of water that was cold, and when he realized there was damage far more severe than he had initially believed and he shut off the power to the control rods and insisted that the reactor had not been damaged. The result was that Akimov directed the men in the turbine room to open the valves of the cooling system which then exposed the valves to a significant amount of radiation.

In the meantime, the people in within the plant could not from hearing the blasts and instantly begin to speculate about what was going on. According to one worker, "At the time of the explosion, I was standing right near the dispatcher's station, which was where I was working. A powerful explosion of steam was then heard. It was not a big deal to us. it, as steam discharges had happened frequently during my time there. I was planning to take a break to sleep, but then there was an explosion. I ran for the window. After the blast, there were more which followed instantaneously...I observed a black flame that flew over the top in the section for turbines in the Unit 4...In the central hall, it appeared like a luminescence or glow. However, there was nothing to ignite, but only the part of the reactor's snout. We determined that the glowing originated from the reactor. ..."

Aerial photos of the damage caused by the explosions

Nikolai Gorbachenko was in the duty room when there was a loud bang. Although he was a bit shocked, he was not worried as he believed it was the result of the turbine located in Unit 4 being closed for testing. But, after hearing an additional sound, he was concerned and was particularly concerned when the lights inside the building were out. He rang the phone and found it also, wasn't functioning. As he walked out of his home at the hallway, he saw that the hallway was suffused of dust and steam, and the needles on radiation counters were flipped around like a flag waving. He later recalled the incident: "At the moment of the explosion, and shortly after I was in the dosimetry monitor. There were several shivers of incredible force. I was thinking: Everything could be a roof. But when I woke up, I was alive, and I was standing up on my feet. A friend of mine, my assistant Pshenichnikov very young man, was alongside me in the dosimetry room. When I opened the door to the corridor that led to the deaerator galleries. Clouds of steam and white dust were coming out of the corridor. The smell was typical of steam. There were sparks of discharges. Short circuits. The panels in Unit 4 were instantly sunk on the panel for dosimetry. There were no readings were recorded. I was not aware of what was happening inside the unit, or what conditions were affecting radiation. Emergency signal systems were functioning within the panel of Unit 3 (we were using a single panel for the entire phase that was being constructed). The instruments had all gone off-scale. I tried to turn off the toggle switch to the control room of the unit but the switchboard had no power."

Everyone at the plant knew about the most serious risk in the event of some kind of accident and, as Gorbachenko stated: "I tried to determine the radiation level in the room I was, as well as the hallway outside of the door. I only had the radiometer DRGZ that was rated for 1,000 microroentgens/second. It was off the scale. There was another instrument that had the scale going upwards of 1,000 roentgens however, when I turned the instrument on as fate was on my side the instrument was out of fuel. There was no other. Then, I went to the control room for the unit and reported the problem to Akimov. It was all over the place at about 1,000 microroentgens/second. It was probably about 4 roentgens an hour. If this is the case it would mean that we could be working for about 5 hours. Depends, of course, on the specifics of the emergency. Akimov advised me to walk around the unit to examine the situation dosimetrically. I went all the way to level of +27 via the staircase-elevator well, but not further. The instrument was off-scale everywhere. Petya Palamarchuk came, and we went into Room 604 in search at Volodya Shashenok... "

Gorbachenko and Palamarchuk Gorbachenko and Palamarchuk discovered Shashenok unconscious inside the room for instruments being held under a beam of a huge size and covered with intense burns resulting from exposure to steam and radiation. They removed him from the structure and directed him to be transferred to the hospital nearest in which he passed away just in a matter of hours.

While Shashenok was unaware that his hands had caused a radioactive mark on his rear, Gorbachenko was looking for Valery Knodemchuk. Knodemchuk was supposed to have been operating the main pump to circulate the air the night before.

Gorbachenko was unable to locate the man, and it was later discovered that Knodemchuk was dead instantly following having been crushed in the debris or evaporated in the aftermath after the explosions. Whatever the cause it was, his body wasn't discovered.

Oleg Genrikh, and Anatoly Kuguz were in a control area at the time of the explosion, and the sound was enough to destroy their window for observation and then put light bulbs out. Both were suffocated by the escape of steam. Genrikh was more severely burned, Kurguz was more seriously burned, while the two joined with two other people in fleeing the building. As they were able to pass Dyatlov on the way out, Dyatlov instructed the men to wash their bodies and to go to the hospital in the plant. Then, Kurguz went by ambulance to the hospital.

Reactor section's foreman Valeriy Ivanovich Perevozchenko was able to see the massive blocks on the top of Upper Biological Shield begin to spin around as the structure started to shake. He rushed down the staircase to the control room in order to warn his colleagues however, by the time his arrival the reactor was exploding. When his skin began to turn to brown in front of him, he attempted to find the people who he recognized were working near the site of the site of the explosion. However, he had to give up after being overcome by weakness and nausea.

Alexander Yuvchenko was in his office at the time of the explosions and he later reported, "There was a heavy impact. After a few seconds I felt a wave enter the room. The concrete walls were bent as if they were rubber. I was convinced that war broke out. We began to search for Knodemchuk...but the man was at the pumps, and had been evaporated. Steam surrounded everything, it was dark, and there was a loud sound of hissing. It was just a ceiling, just the sky, a vast sky of stars. I remember thinking about how stunning that was."

Even though he was burned badly himself, he hurriedly moved through the facility looking for colleagues. He was then joined by three others who were assigned to lower the rods of radioactive control to ensure safety. A massive person, Yuvchenko opened and held the door of the room that was damaged while the other men went into the room to complete their task. Although the three of them were in the room for only one minute, they all were exposed to a fatal dose of radiation. They were dead after just two weeks. To his credit, Yuvchenko reported that though the radiation burns to his shoulders and legs "You do not feel any pain during the time. We didn't realize there was this much radiation. We met a man who had a doseometer , and the needle was right out of the dial. Even then we were thinking , 'Rats! This is an end to our career in nuclear. We all were thinking, 'We've been exposed it's been happening in our time and we began to do whatever we could to help. After approximately one hour I began to vomit inexplicably. My throat felt very painful. We thought we could have had 20or 50rem. However, there was a man there who had been victimized in an nuclear incident in the submarine fleet. the man said it was much more dangerous than the other. "You shouldn't vomit after 50," he explained."

The collapse occurred at about at 6:00 a.m., Yuvchenko was transported to the hospital where he would stay for the remainder of the year.

Chapter 3: The Pressure Of The Invisible Rays

A photo of nuclear fuel spilling out into the underground of the power plant.

"[Valeriy Ivanovich] Perevozchenko...rushed into the corridor...intending to look for his subordinates who could have been in the rubble. One of the first things that he did was go to the windows that were broken and gaze out. The air was filled with a very powerful smell that was fresh and strong, much like the smell of a thunderstorm but ten times more powerful. It was dark in the yard. Reflections of red light from the burning turbine's roof hall in the nighttime low sky. If there was no breeze, the air did not smell. At this moment, Perevozchenko felt as it was the force of unnoticed rays which were flowing through his body. He was overwhelmed by an internal panic that was caused by the demise of his body. However, his fear for his fellow soldiers was the most pressing. He pushed his head away and looked at the right. He noticed that the reactor had been destroyed. The place where the walls of the main pump room were, He saw in the dark the wreckage of pipes, structural elements and other machinery. Then, above...He lifted his head. The drum's spaces separators were also not there. It was a sign of an explosion inside the hall central to. ... The fire was that erupted in his lung. The initial apprehension passed. Perevozchenko experienced a sensation of fire in his chest and on his face, and throughout the entirety of his body. Like he was totally burned out from the inside." -- Grigoriy Medvedev

In the days following that explosion Dyatlov received frequent updates on the damages to the plant, however Dyatlov maintained that it wasn't caused by the reactor's explosion instead of an explosion inside the tank for emergency use. In actual fact, he had stated an "tank blast" in his boss, V.P. Bryukhanov who was present when Bryukhanov was at the plant just an hour after. Dyatlov also ordered Reactor 3 to shut down and instructed Akimov to bring in the day shift to assist with the situation. Dyatlov later met with Gorbachenko and the two began moving around the outside of the plant trying to figure out how much damage was done. When they returned to the room for duty the two men were both physically sick and began to vomit. They were taken to an area hospital for treatment.

Following the time that Dyatlov was taken to hospital, the Chief engineer N.M. Fomin took charge of the situation. initially, he ordered water to be put into what he'd been told was an operating reactor. However, after he had to keep replacing the personnel pushing water into the facility after they had become overwhelmed with radiation sickness, Fomin ordered Anatoliy Sitnikov, who was then an engineer in charge of the deputy in charge of operations, to ascend up to the top of the roof to provide him with an inspection on the condition of the nuclear reactor. Sitnikov returned to Fomin at 10:00 am and advised Fomin that the reactor was in fact destroyed. Although this news could have cost him an entire life Fomin was unable to believe the report and kept the water flowing through the reactor's crumbling structure. The result was to spread radioactive material throughout the damaged pipes and through other buildings on the site.

At this point, the personnel within the plant were in contact with other people in the upper tier of command. They had also been informing them about the incident However, in many cases they had accidentally misled them. Alfa Fedorovna Martynova shared a phone call that she and her husband received that night: "On 26 April 1986 at 3:00 p.m., the intercity telephone rang in our home. Bryukhanov had called [V.V.Maryin from Chernobyl." Maryin who was from Chernobyl. After the discussion was over, Maryin told me: "A terrible accident occurred at Chernobyl however the reactor remains intact. ...' It was quick to get changed into his clothes and demanded his vehicle. As he was leaving he contacted the highest leader for the KGB's Central Committee up through channels. First Frolyshev. He was then referred to as Dolgikh. Dolgikh called Gorbachev as well as the people in the Politburo. Following that, he went to go to his trip to the Central Committee. Around 0800 He called me at home and I was asked to pack his bags to travel."

As people who were on working on the ground within Unit 4 were trying to rectify the situation, regardless of how ill-informed or well-informed they were, a few people who were working in various areas around the nuclear plant feared the most dire scenario and decided to not take any chances. Yuriy Bagdasarov was a shift chief in Unit 3 at the time the incident occurred. it is his fault that he swiftly believed in the worst case scenario and saved his life, and that of his colleagues, by instructing everyone to put on protective gear at the first hint of danger. After not getting the approval of Fomin to turn off the power supply to his reactor, he was able to shut the reactor down nonetheless, thus stopping the spread of contamination of the water flowing from the reactor damaged within Unit 4. This would have likely stopped another catastrophe from happening by preventing melting down of the reactor number 3.

Chapter 4: The Fire Continued To Howl And Then To Howl

"Akimov Toptunov and "Akimov" Toptunov had already ran around the reactor in order to observe the impact on the flow of water from the second feedwater pump. However, the flame continued to growl and to make a loud noise. Akimov Toptunov and Toptunov were already brownish red because of the nuclear sunburn already nausea had irritated their intestines. Dyatlov, Davetbayev, and others from in the turbine room were in the medical facility, they'd already sent shift chief of the unit Vladimir Alekseyevich-Badichev to replace Akimov, but...Akimov and Toptunov weren't leaving. It is only possible to bow one's head in recognition of their courage and bravery. They were, after all, were condemned to a certain end. But, their actions were based on an untrue premise that the reactor was still functioning! They were completely unable to be convinced that the reactor was destroyed and the water was not flowing to it, but carrying nuclear waste along with it and flowing to the levels below, inundating the cables and high-voltage distribution equipment, and posing the risk of stealing the power of all three power-generating units operating." -- Grigoriy Medvedev

After a few minutes, the test failing, the appropriate warnings were sent out and firefighters were there in a matter of minutes to battle the blazes caused by the explosion of the reactor. However, because certain employees did not comprehend the extent that the radiation was causing, even the chief of staff the situation, firefighters entered the scene without the type of clothing that would shield them from radiation and even though many were aware of the dangers they were in, some didn't. Grigorii Khmel was driving an early truck to Chernobyl the next morning. He later said, "We arrived there at 10 or 15 minutes or at least two in the morning...We noticed graphite scattered around. Misha said: Is this graphite?' took it off. However, one of the fighters in the other truck took it. "It's hot," the man said. The graphite fragments were of various sizes and shapes, some large, others smaller, small enough to be able to pick up up...We were not aware of radiation. Even the people who worked in the lab didn't have any idea. It was clear that there was nothing within the vehicles. Misha filled a cistern , and we pointed the water towards the highest point. Then the boys who lost their lives were taken to the roof. Vashchik, Kolya and others and Volodya Pravik...They took to the roof stairs ... as did I. have never seen their bodies again."

A view of fragments of a graphite moderator taken from the central part of the reactor that exploded.

However there were firefighters who knew of the risks of radiation such as Anatoli Zakharov who said, "I remember joking to the other firefighters that there must be a huge amount of radiation around here. We'll be fortunate to be alive in the morning. .'...Of course, we were aware! If we'd abided by the rules that we'd never been close to the reactor. However, it was an obligation of morality - our responsibility. We were like kamikaze."

But their education dominated their thinking and took over, causing them over and over again to risk their lives to stop the blaze that was that was ravaging the plant. They were right where the flames were, some even deep into the inside of the reactor and some of them sacrificed their lives, without realizing the way they were acting. V. V. Bulava who drove one the very first vehicles to Chernobyl plant on the day of the incident was able to recall, "I received an order to place myself under the control under the command of Lt. Khmel. I went to the plant. I put the truck up in the area where the water was, and switched to the source of the water. My truck had been repaired and it was as good as new. It was fresh and smelled like paint. The wheels also had brand new tubes and tires. As I was approaching the car I heard something pound the front fender on the right. I leapt out to check the cause. It was a piece of reinforcement steel had punctured the tire, sticking out of the tire and was catching the fender...It was just repaired and it was a shame. However, for the moment, I had to connect the machine to water, and it was too late. After that, I turned on the pumps, and sat in the cab, however that iron piece was constantly pestering me. I got out , and noticed the iron had punctured the tube, and was reveling. But, I thought, I'm not going to put up with something like that. I climbed out of the truck and began to pull at the thing. It didn't give. It caused me a lot of trouble...And in the end, I ended up being treated in the Moscow clinic with severe burns from radiation on my hands. If I had knew I was going to get burned, I'd put on gloves. That's life..."

So, even though they were exposed to potentially fatal doses of radiation firefighters tried to stop the spread the fire to other reactors inside the plant. But in the end the fire continued to burn through the Unit 4 for two whole weeks until it was finally put out on May 10 , with the aid of nitrogen liquid and helicopters dropping material on the fire from above.

In this image the Major Leonid Telyatnikov, the Commander of the Chernobyl Fire Brigade, is presented with the prize for his work. Telyatnikov endured severe radiation sickness following Chernobyl however, he died of cancer in his 50s.

Chapter 5: Don't Cause Panic

"More than 100 patients had been referred to the medical center. It was the time to act reasonable. But not so fast. Bryukhanov Fomin and Fomin continued"The reactor is in good shape! Put water in it!' deep in his soul Bryukhanov evidently had not forgotten the details of Sitnikov and Solovyev (whose names were changed] and asked for Moscow's approval to disperse Pripyat. However, a clear directive came from Shcherbina who was in contact with his advisor L. P. Drach was in phone contact...Do not provoke anxiety. Then, at the time, Pripyat was a city full made up of nuclear power station employees was getting up. The majority of pupils had gone to school. ...." -- Grigoriy Medvedev

As a catastrophe of unknown size was taking place in the Chernobyl plant in it's early morning of the 26th of April the residents of the nearby area (many who worked at the plant in different shifts, began to rise) and their initial thoughts in the morning were different according to where they resided. 8:30 a.m. would be the scheduled time of each shift change in the day at Chernobyl however the 26th of April would not be like any other. Plant employee Viktor Grigoryevich Smagin explained, "I was to take over the duties of Aleksandr Akimov at 0800 hours in the morning of 26 April 1986. I slept well the night before. I heard no explosions. I woke up at 7:15 am and went to the balcony to take smoke. From my room located on 14th Floor, I am able to see that nuclear power plant. I looked out the direction of the plant and instantly discovered that the hall central of my Unit 4 had been destroyed. Smoke and fire spewed throughout the building. I noticed that everything was decaying. I ran to the phone to contact the unit's the control center, only to find that I discovered that the phones were already disconnected. In order to prevent the information from leaked. I was getting ready to leave. I instructed my wife to shut the windows and doors tightly. Don't let the kids leave the home. It's not her idea to leave the house too. Stay at home and watch TV until I got back...I went out on the street and headed to my bus station. The bus didn't make it towards the facility. Then, they broadcast an announcement via radio saying they wouldn't be going to the second entrance like they normally do, but instead toward Unit 1. The entire area had been taken away from the area by police. The police were not letting anyone in. I then showed them my 24-hour supervisory pass to operating personnel and they let me in however, they did so reluctantly. ... I changed my clothes fast, not knowing at the time that I would soon be back to the medical facility with severe sunburn from nuclear radiation and the dose of around 280 radioactive doses."

However, Lyudmila Kharitonova of Aleksandlrovna, a Senior Engineer at Chernobyl Nuclear Power Plant, recalled the morning differently. Chernobyl Nuclear Power Plant remembers the start of the day in a different manner: "On Saturday, 26 April 1986, all arrangements were made to celebrate this May Day holiday. It was a warm day. It was the time of spring. The gardens were in bloom. My husband, who was the chief of the section responsible for the adjustments to the ventilation, planned to take the kids to the dacha following work. I was washing the dishes since the morning and was hanging up my bedding across the deck. Even in the evening there were millions of particles built up on it. In the majority of employees working in the construction industry, nobody was aware of anything. After that, there was a leak about the incident and the fire in Unit 4. What actually happened nobody really knew. Children went to school, the children played on the streets in sandboxes and on their bikes. In the evening of the 26th of April the radioactivity that was present in their hair and clothing of everyone was already elevated, but we were unaware of it when we were there. On our street, just a few meters away , they were selling delicious doughnuts. It was just a typical holiday day."

Although many of those who observed the destruction from afar Some thought there was nothing to worry about. Grigoriy Medvedev provided an account of how things was still quite normal for fishermen despite being near the facility "Fishermen seem to be able to swap each other throughout the night at the location at the point where the drain drained into the cooling pond. Everyone caught fish even when they were not working. It was warm after passing through the turbines as well as the heat exchange devices, which means there were many bites. It was also springand spawning time and fishing was superb. It's around 2km from the place where you fish up to unit 4. ... After they witnessed the explosions and the flames, many stayed at the fishing spot until dawn and others, experiencing unfathomable anxiety and a sudden dryness in their throats and burning within their eyes walked towards Pripyat. The population had grown accustomed to not to pay attention to sounds like the booms like cannons when safety valves were operating and sounded like explosions however the fire was not. .... They would then put it out. It's not much!"

Naturally, the people arriving at the plant were soon to realize that something was horribly wrong just a few hours prior and, while some reacted with fear, others acted in a heroic manner. Others, such as Viktor Smagin, were simply eager to discover the situation and what they saw is chaos. The pressure of the situation prompting various arguments that were short-lived. Smagin wrote "I were in a rush so I put on my cotton coveralls, high boots and the cap "respirator 200" and ran down the long corridor of deaerator galleries (which linked the four units) towards Unit 4. In the room where was the computer known as 'Skala' there was a crack where water had dripped from the ceiling to the cabinets that housed the equipment. At the time I didn't realize it was radioactive. There was no one else in the room. ... Then I moved further. Krasnozhon the director of the deputy department of department for radiation safety was working in the room that contained the dosimetry monitor. ... Samoylenko, head of the dosimetrists working in the night shift was also there. ... Samoylenko was insisting that radiation was huge and that the radiation was massive, while Krasnozhon was claiming that it was possible to work for 5 hours on 25 per cent. "How long do you have to be working, guys?' I inquired, interspersing their conversation. The background is 1000 microroentgens a second, that is 3.6 ROENTGENS per hour. It is possible to work for five hours, assuming you're earning 25 remunerations!' "It's all lies," Samoylenko summarized. Krasnozhon was enraged over and over and over."

Alongside the emotionally charged atmosphere which caused a variety of anger The radioactively charged environment had its own consequences for people who had been exposed to high doses radiation were experiencing a variety reaction to physical symptoms, ranging from vomiting to panic. Smagin went on to say, "All the windows in the corridors of dearator galleries were destroyed by the blast. There was a strong, acrid smell of Ozone. My body felt the intense radiation. However, they claim that there aren't any sense organs. It is evident that there is something. There was a painful sensation in the chest. It was a vague feeling of anxiety However, I managed myself and held my hand. The air was already a bit light, and the heap of rubble was clearly visible through the window. There was something black strewn all over across the pavement. I looked it up and realized it was probably reactor graphite! This isn't bad! I could see how the reaction was operating in a negative way. However, the real-life consequences of what had occurred was not clear to me."

As the account of Smagin reveals the hysteria that was taking place in the plant and the panicked administration were unable to decide what they should do the next step. Vladimir Pavlovich Voloshko, the chairman of the Communist Party Committee in Pripyat was in the plant when the incident occurred and was able to describe the chaotic circumstances in Chernobyl: "The entire day of the 26th of April Bryukhanov was a mess like a person who was lost. Fomin the day, he would cry during pauses between issuance of orders. He was losing confidence in himself. Both mostly resigned their senses by the evening, at they had arrived at the moment Shcherbina arrived. As if he could have brought peace to them. him...They sent Sitnikov the brilliant scientist, to consume 1,500 Roentgens! They didn't pay attention to his report when he announced that the reactor was destroyed. Of the 5,500 workers employed by the plant 4000 disappeared to places unidentified on the very initial date... "

In actual fact it was clear that there was no longer anything to be done to save the demolished Unit 4, a realization which managers could have made within minutes of the explosions, not hours later. Whatever the case, it was time to devote their efforts towards rescuing the people inside the facility and those living in the vicinity, who's lives were put at risk with each passing minute that information was not shared. It was not surprising that the power plant was locked down in place that ensured that no one from the power plant could go home after the conclusion of their shift and this worried the families of workers. Lyubov Nikolayevna Akimovawho was spouse of Unit 4 chief Aleksandr Akimov recalls "The whole first portion of the day, I was running around, asking everybody, and searched to find my husband. Everyone knew there was an accident, but I was overwhelmed by more anxiety. My journey took me to Voloshko at the gorispolkom as well as in the party gorkom, and to Gamanyuk at the Party Gorkom. Then, after asking several people, I found out that he was in the medical center. I ran over. However, they refused to allow me in. They claimed that he was getting the intravenous infusion at that moment. I did not leave. I went to the window in his room. He soon came to the window. His face was dark reddish brown. As he saw me, his face began to light up at me. He was thrilled, and he assured me, asking what I thought of the sons via the glass. It appeared to me that at this time, he was grateful to have boys. He advised me to not allow them to go out on the streets. He seemed even happy and I felt at ease." Akimov passed away two weeks later due to the radiation poisoning which he contracted that day.

A growing number of people within the vicinity of the Chernobyl power plant realized that something was happening in the area as time passed on April 26, however, nobody was aware of how important the situation was and they didn't have a reliable source of information either. Nadezhda Petrovna Vygovskaya resided in the village in Pripyat with her son and husband. She wrote about it on the weekend: "That morning no one thought anything was wrong. My son went to school, and my spouse went out to the hairdresser's. I'm making dinner when my husband gets back. "There's a burning at the nuclear facility He states. They're telling us not allowed to turn on the radio.' ... I still can be able to see the bright red glow that was as if the reactor was shining. It wasn't just a normal fire, it was a kind of shining. It was beautiful. I'd never seen anything quite like it in a movie. In the evening, everyone poured onto their balconies and those that didn't have the space went to their the homes of their friends. We were located on the ninth floor and we had a fantastic view. Kids were brought out by parents to pick them up and told them, "Look! Remember!' They were the people who worked at the reactor: engineers or workers, physics teachers, and even physics workers. They sat in the black dust and talked, breathed and pondering the thing. People came from all over in their cars and bicycles to look. We didn't realize the possibility of death being this beautiful. Although I'm not saying that there was no smellit was not a spring or autumn scent, but something different and it wasn't earthy. My throat twitched and tears began to fill my eyes."

The 26th of July, Lyudmila Kharitonova thought that it was better to move her family out of the region for a few days and explained "We chose to go to The dacha (a country home) however there were police stationed along the road and refused to allow us to leave town. We returned to our to our home. Strangely, we thought of the incident as distinct from our personal lives. In all likelihood, there were accidents in the past however, they were only affecting those who worked at the facility itself...After dinner, the workers started to clean all the roads of town. However, this didn't draw the attention of anyone. It was just something that happened in the summer heat. Sprinkler trucks are nothing uncommon in summer. It was a normal, peaceful setting. While I did glance at the white foam that was forming in the drains, I did not pay any attention. I believed that the pressure of the water was very high. Children from the neighborhood were cycling on the overpass. From there you could get a clear perspective of the unit. the incident had taken place in towards Yanov Station. Then we learned that it was the radioactive spot in the city, as the radioactive discharge cloud was passing through there. It was later discovered when, in the morning of the 26th of April the kids were curious about the burning reactor. Those children later developed serious radiation disease."

In the end, news was spread that there was an accident that was serious, however the government did not issue an the order to evacuate. In fact the way Kharitonova remembers, "After lunch, our children returned from school. They had been told not to walk out in the street to do the house cleaning with water, which was when the public first realized it was a serious issue. People learned of the incident at various moments, but by night of April 26 everyone was aware, however the response was calm as all the shops as well as schools and institutions were in operation. We believed that this indicated that the incident was not as risky. The odor became more alarming as the evening drew closer. The feeling of unease spread from unknown sources, possibly from within the soul or due to the atmosphere in which the metallic smell had grown intense. The exact cause I am not able to be sure. However, it was metallic... The evening time the fire was much more than usual. The graphite had started to burn. The fire was visible from afar but did not pay much focus. "Something has caught fire ...' The firemen have taken it out ....' It remains burning .'..."

The plant began to think about what to do the next step, a shift in the leadership was observed from those with expertise in the science of the operation of the plant to those with the experience and knowledge needed to manage what was to come next. Grigoriy Medvedev detailed just part of the command chain through his memoir: "At the beginning of the eighties the nuclear power sector was managed by the Central Committee, Maryin headed it, and eventually it was staffed by assistants. G.A. Shasharin G.A. Shasharin, a seasoned nuclear engineer who worked for many years at operating nuclear power plants, and will in the future serve as the deputy minister of power in charge of operating nuclear power plants and nuclear power plants, was among the group. With him, Maryin was traveling to the unit that was damaged in Kizima's GAZ vehicle. In the process they encountered private vehicles and buses. The spontaneous evacuation had started. Many had fled Pripyat for good, even on the morning of April 26 together with family members and their radioactive possessions in the absence of waiting for instructions of the local authorities."

G.A. Shasharin was an engineer. He now brings his organizational and structural expertise into the problem. He later explained,

"On the journey between Kiev on the way to Pripyat, I told [A.I.Mayorets] Mayorets about the working group. I was thinking about this prior to my flight between Simferopol towards Kiev. Here is a list of groups I had suggested:

1.) A group of experts to investigate the root causes of the accident and also the safety of the plant ....

2.) an organization to research the radiation levels surrounding the nuclear power plant ....

3) an organization to repair the damages and restart operations ....

4.) an evaluation group to determine whether it is necessary to expel the inhabitants of Pripyat as well as nearby villages and farms ....

5) an organization that provides instruments as well as equipment and supplies ..."

But, when Shasharin was formulating his plan for dealing to the situation, a lot of the people around him were slow to respond unwilling to lead the way and consequently shoulder the burden and blame for the choices that were later made. Vladimir Nikolayevich Shishkin was the deputy chief of Soyuzelektromontazh of USSR Minenergo, and he was in the discussion regarding what to do with Pripyat. He said, "It seemed that all people responsible for the disaster were looking to delay the day of complete acknowledgment, at which point all the i's will be crossed. They desired, as was commonly prior to Chernobyl the accountability and blame to be spread out slowly over all. This was the main reason behind the delay, when each second was precious, and when delay could threaten the innocent citizens with radiation. It was clear on everyone's minds the word "evacuation" was hitting the people's skulls...But it was the case that the nuclear reactor was on fire all the time. The graphite was burningand spewing thousands of curies of radioactivity into the air. Despite the tense and potentially dangerous situation in the damaged facility the atmosphere in Pripyat is calm and businesslike Gamanyuk, the first director of Pripyat Party Gorkom and Mayorets, said. to the Mayorets (at the time of the accident , he had been in the medical center for an exam, but at the dawn of April 26 he left in the medical unit and went back to work). "There was no chaos or panic. A normal, everyday life on a holiday. Children are playing on the streets, sporting events are happening, classes at schools. Weddings are also being celebrated. Today, they've been celebrating 16 weddings of Komsomol youngsters. We stopped fake rumors and gossip."

The people in charge believed there was no need to evacuate anyone if they could simply prevent radioactive material from going out of the facility. Shasharin recalls, "Later, we went up in the helicopter along with Maryin as well as the vice chairman of Gosatomenergonadzor and Sidorenko who was a corresponding members of Sidorenko, a member of the USSR Academy of Sciences. We flew over the helicopter at an altitude between 250 and 300 meters. It appears that the pilot was carrying an dosimeter. However, there was there was no radiometer. At that height it was a radiation of 300 roentgens/hour. The upper slab was warmed up to an intense yellow shade which was in contrast to the bright cherry hue reported by Prushinskiy. That means that the temperature inside the reactor was increasing. This slab wasn't as straight when it was placed on the shaft. It was as straight as later on, when they put into baggies of sand. The weight moved it around. It been made clear that the reactor had been destroyed. Sidorenko had proposed to throw around 40 ton of lead inside the reactor to lessen radiation. I strongly opposed it. This kind of weight at an altitude of around 200 meters was a massive dynamic load. It would create an opening all the way up to the pond with bubbles and the whole melted core would be drained into the pond's water. You would then be forced to run where your legs will lead you."

Although the fire's ongoing blaze was a attraction throughout the entire day of April 26th, the fact that the flames were still burning at the time it got dark led many people who lived nearby to believe that something was very wrong. Nadezhda Vygovskaya wrote "I was awake all nightand heard the people in the neighborhood wandering around upstairs, not sleeping. They were carrying things about, banging on things or maybe you could have seen them packing away their possessions. I battled my headache by taking Citramon tablets. When I woke up in the morning, I looked up, looked around, and I was feeling it wasn't something I thought of making up, I was thinking that at the time it was like something wasn't quite right and something is different. When I woke up at 8 am, there were military personnel out in the streets wearing gas masks. When we saw them in the streets, accompanied by all their military equipment, we did not be scared -- on contrary it calmed us. Because the army is coming to help us it's all good. We didn't know at the time that this peaceful particle can cause death, and that mankind is powerless in the face of the laws of physical physics. Through the day, on the radio, they were warning people to be prepared for an evacuation. They'd leave us in three days. clean everything, then check it. The children were instructed to pack their schoolbooks. However, my husband tucked the wedding papers and documents photographs into his briefcase. My only item was a gauze-lined kerchief in the event that weather got bad."

Chapter 6: Come On, Come On

Image of a radioactivity warning sign in Pripyat

"He was an early chairman on the government's commission to assess the damages of the nuclear accident at Chernobyl. A little more pale than normal with his lips squeezed tight and the fierce appearance at the folds of his cheeks, he seemed serene, composed, and focused. At this point the man didn't realize that all around in the streets and inside the air was filled with radioactivity, emitting beta and gamma rays and was completely indifferent to who was being affected--the of the devil's own, ministers or common mortals. He was blessed with tremendous power, however, it was still a human being and it was taking the same path humans go through in being First the storm would begin to build beneath the backdrop of calm outside, and after he'd discovered something and was laying out the strategy that would be used, the actual storm would break out in a ferocious rush and impatience speed, speed! "Come on! Come on!" - Grigoriy Medvedev

In the end, the decision to evacuate areas around was finally taken. Shcherbina told his colleagues in the administration "We are moving out of the city in the early morning of 27 April. Get all 1,100 buses in the night and transfer them to the highway that runs between Chernobyl as well as Pripyat. General Bedrov must install sentries at each home. There should be no one on the streets. The next morning, civil defense will broadcast the required information to the public over the radio. Also, they will announce the exact date for the evacuation. Bring potassium iodide tablets to those living at home. Request Komsomol members for this purpose."

With these instructions in place an excavation order was sent to residents from Pripyat: "For the attention of all residents of Pripyat! Pripyat City Council informs you that because of the accident at the Chernobyl Power Station located in Pripyat, the capital city. Pripyat the conditions of radioactivity within the area are becoming worse. It is the Communist Party, its officials and the army are taking the necessary measures to stop this. But, in order to ensure that the population is as secure and healthy as they can and with children as the top priority, we must temporarily exile citizens living in the closest towns of Kiev Oblast. This is why, starting on April 27, 1986, at 2pm, each apartment block will be served by a bus at its disposal, which will be supervised by the police and city officials. It is recommended to carry your documents, important personal items, and a small amount of food in case you need to carry it. The chief executives of the public and industrial facilities in Pripyat have made the employees who will need to remain in Pripyat to ensure that these facilities are in good condition. All residences will be secured by police throughout the period of evacuation. If you are leaving your home temporarily, please ensure that you've shut off the electric equipment, lights as well as the water supply and shut the windows. Be calm and organized during the short-term evacuation."

This is how the inhabitants In Pripyat unexpectedly found themselves on hundreds of buses that would take them away from their homes, maybe for ever, with no idea of what they were taking. G.N. Petrov was one of the people who left the following morning "At approximately 1400 pm The buses were at all entrances. We were warned on the radiothat we should dress casually and take only a few items, we'd return in three days. However, the uncontrollable idea popped up: if a lot of things were taken, even 1,000 buses wouldn't suffice. The majority of people obeyed and didn't even use any money. However, our people are nice They joked and laughed with one another, and they encouraged the kids. They would tell them: we're going to visit grandma...to the festival of films festival...to the circus. .... The children and the adults were sad, pale, and mute. Anxiety and forceful joy were in the air , along as the radiation. However, it was all effective. A lot of people had gone downstairs prior to time and were crowded outside with their kids. They would keep asking for them to return to the entrance. As soon as they announced the departure, we stepped out of the gate and straight to the vehicle. People who stayed behind ran from bus to bus just to take rems that were not needed. So, during the course of a peaceful normal life, we'd taken plenty of money both outside both inside and outside."

Alexey Akindinov's "Chernobyl. The last date of the Pripyat"

A photograph from 2001 of an abandoned house close to Pripyat

Many fled Pripyat on that day, each scared, worried about their well-being, their futures, and the future of relatives. They had a longand difficult journey ahead and it would go on for a long time after the actual journey ended. But, a lot of them were up to the challenge and as Nadezhda Vygovskaya said: "As we were leaving Pripyat there was an army line that was heading back in the opposite direction. There were several military vehicles that I was scared. However, I could not forget that it happened to somebody else. My tears were flowing, searching for food, lying down with my son, calming him yet inside there was this feeling of being a spectator. In Kiev they offered us a little cash, but we weren't able to purchase anything. Hundreds of thousands were removed and had bought everything they could and devoured all the food they could. Many suffered strokes and heart attacks in the train stationsor on buses. My life was saved thanks to my mom. She'd been a very long-lived woman and had gone through everything at least once. ... Then she declared, "We have to get through this. In the end, we're alive. I'm able to recall one thing: on the bus and everyone is crying. The man in front is screaming at his wife. "I cannot believe you're such a fool! Everyone else has brought their belongings and all we have is 3-liter containers!" The wife decided that, as they were taking their bus ride, they could like to take pickling bottles that were empty to her mother, who was traveling with them. They had huge bulky bags on their seats. We were running over them all the journey to Kiev and that's exactly what they went to Kiev with."

Of course one of the challenges that the people who coordinated the evacuation had to face was the decision of where to place refugees, particularly since several of the villages in this area already overflowing with people who had already left prior to the official evacuation going into force. G.N. Petrov noted, "We drove to Ivankov (60 kilometers from Pripyat) and found ourselves scattered across the villages. Some of them did not accept us without hesitation. One wealthy peasant wouldn't allow my family to enter his huge brick home, not because of the risk of radiation (he didn't know what was going on and explanations did not have any impact on him) however, it was due to desire to be greedy. The house was not constructed for the purpose of being able to allow strangers into it. .... Many of the people who had quit on the busses in Ivankov continued on to Kiev on walking. A few people took were able to get rides on a hitch. A pilot from a helicopter I know I was able to hear later what he'd seen from the sky: massive crowds of people in light clothing including children and mothers and elderly people walking on the roads and shoulders towards Kiev. He saw them in the vicinity that is Irpen as well as Brovary. The vehicles were snared in the crowds just like they would be during the course of a cattle drive. It is common to see scenes like this in films from Central Asia, and you immediately think of a similarity even if it's a poor one. Trudging and trudging... "

If you are not living in the area the state-controlled TV station broadcasted to other viewers across in the Soviet Union on April 28, "There has been an accident at the Chernobyl Nuclear Power Plant. One of the reactors suffered damage. The effects of the incident are being dealt with. Assistance is being provided to anyone affected. An investigation commission was established." But the subtle manner of the news was not able to fool anyone and many were hesitant to be in contact with people who had been exposed to radiation. Nadezhda Vygovskaya remarked, "From the very first I was convinced I was Chernobylite, being an individual group. The bus we were on stopped at night in a village. there were people sleeping on the floor in a school, and others in an organization. There was no place to go. One woman offered us to stay at her home. "Come,' she told us she would lay down some bedding for you. I'm sorry for your son. Her friend began taking off from the group. 'Are you crazy? You're contaminated!' As we arrived in Mogilev after our kid began school, he returned on the first day of school with tears. He was placed in a class with one girl who claimed she didn't want to be with himas being radioactive. My son had been in fourth grade and was the only child who was from Chernobyl that was in class. Other kids were scared of him. They called him "Shiny. His life was over so quickly."

Chapter 7: Required Bags, Shovels, Sand And People

"Gen Antoshkin gave up his spot on the roof of the hotel 'Pripyat' to Colonel Nesterov in order that he could guide the light and he himself would rise in the air. For a long period, he was unable to determine where the reactor was located. If you weren't familiar of the structure of the unit you were unable to find your orientation. He was aware that skilled assembly personnel or operating personnel were required to accompany the 'bomb runs' '...The reconnoitering was completed before the flight paths for the unit were identified. They required bags with shovels, sand, and people to take the bags, fill them with sand, and drive them to the helicopters. General Antoshkin spoke about the details to Shcherbina. Everyone in the group gorkom had coughed, their throats dry and they had trouble speaking." Grigoriy Medvedev

While the authorities in charge had finally issued an evacuation of the reactor, it continued to emit radiation at a rapid rate and needed to be shut down. At this point the General N.T. Antoshkin took over. As per G.A. Shasharin, "Air Force Gen Antoshkin did very good work. A feisty and professional general, he never gave anyone peace, and he snarled the entire group. Around 500 meters away from the gorkom, close to the "Pripyat" Cafe they discovered a heap of fine sand close to the river's terminus. Hydraulic dredges made it to be used for construction of the new housing areas in the city. They brought bags out of the storage area at the division for worker supply and we, the three of us, I, A.G. Meshkov, the first vice minister for medium-sized machine construction Gen Antoshkin and A.G. Meshkov began to load the bags. Soon we were sweating. We did our work as we had been: Meshkov and I in Moscow suits as well as street shoes and the general dressed in his uniform. We were all without respirators or dosimeters. I soon involved in this effort Antonshchuk, manager of the trust Yuzhatomenergomontazh, his chief engineer A.I. Zayats, Yu.N. Bypiraylo the administration chief of GEM and many others. Antonshchuk came up to me with the benefits list, which I found funny, but I endorsed the offer immediately. Antonshchuk as well as those planning to get to work followed the old system without realizing that the "dirty" zone was everywhere and the benefits needed to be given to every person who was living within the urban area. I didn't intend to disorient people by giving explanations. There was work to be completed. However, there weren't enough people in the area. I demanded A.I. Zayats, chief engineer of Yuzhatomenergomontazh, to go to nearby kolkhozes and ask for help...."

It's difficult to imagine a situation like the one that saw three men pouring sand in the massive reactor manually Then it became apparent that they weren't up to the task to the task and needed help. Of course, it would be challenging to find suitable volunteers for this kind of job, and Anatoliy Ivanovich Zayats explained the problems: "Antonshchuk and I went to the farms in the Druzhba Kolkhoz. We walked between farms. People were working in the plots that were around the house. Many were out on the field. It was springand planting time. We started to explain that the soil was not suitable as well as in that the melter's throat needed been plugged and we required assistance. It was very hot since the beginning of the day. The people seemed to be in the Sunday pre-holiday mood. They were skeptical of us. They continued to work. Then we came across the kolkhoz the chairman and secretary of the party's governing body. They went to the fields along with us. We explained the process over and over. Then, the audience were able to understand us. A total of 150 volunteers came together, men and women. They then began to load the bags onto the helicopters with no breaks. All this was carried out without breathing masks or other safety equipment. On the 27th of April they facilitated 110 helicopter flights. On 28 April 300 flights were supported by helicopters. ...."

In the end, this is what happened when the residents of Chernobyl have filled the reactor's belching with sand. By doing this, they probably endangered and even protected their lives, because although they were exposed to greater radiation that they might have in the absence of their help but they were exposed to lesser radiation than what they could have experienced had nobody had intervened.

Chapter 8: The Dozens Of People Who Died

"I am thinking of the many victims, those whom we know names as well as the many in the womb, their lives disrupted by the names of those we'll never be able to learn because they perished because of the an interruption in the pregnant women who had been irradiated at Pripyat on the 26th and 27th of April. ... They were firemen and operators. The doctors continue fighting to save the lives of others, both serious and less seriously ill patients. Staff staff from the headquarters of USSR Minenergo have kept watch in the clinic, supporting medical personnel." Grigoriy Medvedev

A large portion of the people injured in the blast at Chernobyl had their final days in one of the numerous hospitals located near the power plant. However, certain men survived their injuries and exposure . They survived to tell the experience of the medical professionals who were overwhelmed to put their lives at risk in order to save others. V.G. Smagin V.G. Smagin, whose shift in the factory was the same as Akimov's shift, told of, "They sat the five of us in an ambulance and drove us to the medical facility in Pripyat. They utilized the RUP to test everyone's radioactivity. We washed a few times. It was still radioactive. There were a number of therapy assistants in the room with doctors. Lyudmila Ivanovna Prilepskaya immediately took me away to her office. Her husband is also the shift chief. Both of our families were welcoming. However, at this point me and the other guys started vomiting. Then we saw what appeared to be a bucket, or the wastebasket. We took it , and three of us simultaneously started vomiting into that bucket. Prilepskaya inquired where I'd been and what radiation fields were? She could not comprehend that there were fields everywhere, and it was filthy everywhere. The whole nuclear plant was a constant radiation field. I shared with her what I could in between bouts vomiting. I explained that we didn't know precisely what the fields were. They were off the scale at 1,000 microroentgens/second and that was it."

The good news is that Smagin had less trauma than his fellow comrades as well as, despite receiving treatment himself and receiving treatment, he managed to inspire those in his vicinity. Smagin explained "They placed the IV needle into my vein. Within two hours I started to feel more vigor and energy throughout my body. After the IV was completed I stood up and started looking for a cigarette. There were two other people in the Ward. One bed was occupied by a guard was an Ensign. He was constantly saying"I'm going to escape from here and return to home. My children and wife are scared. They aren't sure where I am. And I don't know what been happening in their lives.' 'Just lie still,' I told him. "You've earned rems. now you can get treatment. ...' The other bed was an apprentice adjuster of the Chernobyl startup and adjustment company. He was informed that Volodya Shashenok had passed away early in the morning, which was at about 0600 and he started to yell at them for having not told him that he had died Why hadn't they informed him? He became hysterical. He seemed to be terrified. If Shashenok had passed away, that means that he too could pass away. He yelled loudly"They're hiding everything, burying everything. ...! Why didn't they let me know what they were hiding? He calmed down however, he was then struck with a tiring episode of bleeding."

The funny aspect of Pripyat's hospital Pripyat could be that it was always in threat of being a radioactive hotspot in and of itself. In the end, as Smagin pointed out, "It was 'dirty' inside the medical center. The device showed radioactivity. They had mobilized women from Yuzhatomenergomontazh. They were always washing the floors of the corridor as well as in the Wards. Dosimetrists would arrive to measure every single thing. He would constantly mutter to himself every time: "They wash and wash, but the clothes are still filthy... '"

Yuvchenko who kept the door open for three other people to try the impossible, and fatally deadly task of dismantling the rods of radioactivity that controlled the operation who were in the hospital suffering from radiation exposure that could easily caused death. In the first few hours of his stay in the medical facility, there was an interview conducted by a nurses and 3 KGB officials, and when the decision was made that he was well enough for travel escorted via bus to Moscow for specialized treatment. Five of the people traveling along with him died just minutes after arriving in the city. when he reached the hospital, the staff cut his hair and placed the patient in an area called a ward. In the next few days, he began experience a sour but not productive cough as well as skin rashes and lips, as well as severe episodes of vomiting and nausea. As time passed, the vomiting diminished, but by the time the skin cells begun to die quicker than the body could replace them. After a while it was a while before he could wake up to a black pile of dust that fell off his body in the previous night. Then, he suffered deep tissue injuries on the areas of his body that been holding open the door of the room that housed radiation rods. In reality, the wounds were so severe in some areas that he almost lost one of his severely muscles-bound arms. Although he was able endure the injury the wound, the wound took 7 years to heal. Then, he had to spend the rest living in a hospital as well as a rehabilitation center however, he was positive enough to tell jokes at times, "The doctors told me that if you've gotten through this, you shouldn't be worried about the rest of your life."

Of course, not all were lucky enough. Lyudmilla Ignatenko was devastated when her husband passed away, Vasily One of the crews assigned to battle the fire in the plant.

She went on to talk about his final days: "At the morgue they asked, 'Would you like to see how we dress him up in?' And I did! They dressed him in formal attire, and wore caps for his services. The shoes wouldn't fit his feet due to his feet having swelling. The doctors had to slice the formal attire to fit him, also because they could not put it on him. There was no body to wear it. All they had were injuries. The two days I was in the hospital I'd raise his arm and the bone was shaking, and it's hanging there, the body had disappeared from it. Lungs fragments as well as his liver were emanating from his mouth. He was choked on the organs of his body. I would wrap my hands in a bandage, then put it inside his mouth. remove all the stuff. It's not something you can talk about. It's not something you can write about. Even to experience. It was entirely my fault."

Chapter 9: The Ash Fell On Pripyat

"Toward the evening of 9 May, around 3000 hours of the night, portion of the graphite within the reactor started to burn, an void was formed beneath the weight that was removed, and the huge pile of 5,000 tons of clay, sand and boron carbide crashed down, throwing an enormous quantity of nuclear ash from under the pile. The radioactivity increased dramatically at the site, in Pripyat in the 30-km zone. The increased radioactivity was also felt at 60km away in Ivankov as well as other areas. The darkness was already beginning to set in they climbed up the helicopter in a frightful manner and determined the amount of radioactivity. The ash fell upon Pripyat and the surrounding fields." Grigoriy Medvedev

When the situation in the plant was stabilized it was the time that the Russian government started sending in groups comprised of "liquidators" to clear the radioactive waste left behind. The first year following the plant's meltdown over 200,000 workers were brought into the plant and one of them was an individual known as Sergei B. He said, "My brigade was headquartered in the nearby village of Oranoe. I was convinced that I had the knowledge and expertise to complete the task and I was able to complete the benefit of my second (military) training in the field of officer-chemist. In 1986 Ph.D. scientist had "carte blanche," meaning they were able to reject any draft request (army reserve) between May and June 1986, there was a huge call-on of mid-rank officers due to the high number of personnel being rotated in the initial "liquidation" campaign. Ph.D. scientist were considered to be a sort like "untouchables". However, I did volunteer. I was thirty years old. I was a lecturer/scientist at the Institute of Chemical Technology, Dnepropetrovsk. I didn't have a idea of what was happening when I arrived. After the first few weeks, it was difficult to gain a clear picture, and a clear understanding of the scope of operations and the amount of money spent in repairs. In the early part of August, I realized that this was more than a single country issue It is a global issue ..."

Although many of the workers were military personnel however, some were civilians who were coerced with threat and assurances to work at the plant that was destroyed. Natalia

Manzurova was willing to help clean up the mess however she regretted the decision she made "I didn't know what I was doing that I was in the wrong and didn't know the full magnitude of the situation until later. The entire incident was covered up with the shadows. I went to the site as a professional since I was instructed to do so -However, if required to investigate an incident now I'd never be willing to accept it. ... The scene was as if I was in a war zone, where a neutron blast had been detonated. It was like being at the center of war in which the enemy was invisibly. The entire buildings and houses were still intact, including all furniture, however there wasn't one person left. All was silence. Sometimes , I felt as if I was the only one living on this strange planet. There are no words to describe the feeling."

The majority of the employees were extremely proud of the work they had done and, as Sergei said, "It wasn't a feeling of doomsday. It was a feeling of satisfaction with what we're about to accomplish, and a realization that the job must be done, no matter what. A huge boost in confidence. It helped diminish the "I'm putting the life of my family here' sense." But, the optimism didn't make the work any less hazardous the fact that Sergei knew well: "Time on the roof was varying from 45 seconds. to three minutes according to the current radiation levels and the location you had to work. Sometime, especially after the helicopter's treatment (they utilized special methods to reduce dust and radiation by dropping tonnes and tons of deactivating solution early in the morning, prior to us beginning working in the area) the levels weren't as high, and could take a little longer. We cut asphalt that contained highly radioactive solids that were melted into asphalt on the day of the explosion (the asphalt had solidified following the initial fire laid down) ...) and then tossed over the ground and over the roof edge. The two raids that I was part of mostly supervised my troops as a result of the fact that my total dose was already high , and I wasn't permitted to take greater than 1.5 Roentgen per trip... this was a shame. We carried parts of the General Military Protective Kit (boots and gloves, head gear, ...), an industrial respirator, as well as a distinctive "protective piece" comprised of 2 thin (about 1/8 - 1/4 inches.) rectangular lead plates that measured roughly 1.5 by 2.0 feet, that covered both sides. They were tacked together."

Due to his rank and experience Sergei worked at Chernobyl longer than many of the other liquidators this meant that he was subjected to greater radiation levels than many of the liquidators. Because of this the man saw how often workers were shuffled between shifts: "I had to lead and be accountable for the safe and effective operation of a troop squad that ranged from 10-25 soldiers (army reserveists). The movement of the troops was awe-inspiring: there was no single day and I was able to make fifteen trips in the row over the course of the month of August... When I had more than three people who were with me the day before me. I couldn't remember their names, and I was unable to remember their faces due to the fact that most times we were wearing respirators or masks, making it really difficult to recognize someone. ...I was part of the major clean-up activities that ranged from the top of the 3rd reactor (highest radiation levels) to the corridors for 1+2 reactor clean-up (lowest likely in the moment). There was also a need to construct a barb-wire fence around the entire station in order to improve the security level. Another tale."

In the end, the people who helped clean up Chernobyl were as heroes and victims of the tragedy, as many of them were sick with the same diseases they had worked to keep others from. In the meantime the workers were exposed to higher levels of radiation than they were willing to admit. As per Sergei, "I had accumulated the holy number of 25 Roentgens. In reality my dose was at least trice higher (according to my estimates) - during my time in the Zone, we did not have individual dosimeters whatsoever, the dose was calculated based on the 'average' working irradiation measurements, 6-8 checkpoints on the perimeter/mid-section of the operational field, surface, etc. Based on the level, the team leader , like myself, established the average time for workby calculating the daily doses that was 2.5 Roentgens (not greater) in each individual. ...there were not enough individual dosimeters in the spring and summer of 1986 for everyone Only 'civics' scientists and engineers were equipped with the devices. We. "military," the army reservists did not. "Shortage" was a term that was commonly used. Another reason was that it was not enough subs/rotation available, particularly of mid-rank officers, so doses accumulated by many of us during July-August-September of 1986 were artificially lowered in the paperwork."

A Soviet medal given to "liquidators"

To make matters even more difficult The secrecy surrounding the Chernobyl catastrophe was so fervent that the government even resisted paying some of the people involved in cleanup directly for their efforts. Sergei said, "I was paid after I received my money back, but oddly, it was a remuneration from my Institute and not the government. Ukrainian Government took care of the cost of my pension, which wasn't so big following my initial substantial salary following the collapse of the USSR was in collapse (initial payment was around five times my monthly income for the 3+ months that I was living in Chernobyl... This was a specific bonus for work in a hazardous environment. ')... that was quite a pity to get my pension. After the first year, people lost interest in who we were and the work we performed to help their benefit... They became tolerant about the (liquidator's) advantages. I've heard plenty of yelling in my back, when I was using a certificate, which I used very rarely anyway...I was diagnosed with some health problems (heart kidneys, kidneys, etc.), which were somewhat contained not because of the government help, but because of my connections/networking - I have several friends among prominent doctors back home.... Thus, I was fortunate to be able to recover without having to go to the hospital, and went on to continue working for my Institute as an Associate Professor. ... It was was extremely cautious and didn't do any foolish things such as eating food that was contaminated with fruits and vegetables as some of my fellow liquidators did during"the Zone."

In the state broadcast, and the issue of the pay of workers became painfully evident the Soviet government was determined to avoid blaming the entire severity of the catastrophe as well as the health issues that people as well as citizens in the aftermath developed. One of the victims was cleaner Natalia Manzurova, who later suffered from thyroid cancer and spoke of the debilitating consequences: "I started to feel like I was suffering from the flu. I would have an extreme temperature, and then start to shake. The first time you come in encounter with radiation? Your healthy health flora becomes depleted and your bad flora is able to thrive. I began to feel like I wanted to rest throughout the day and eat lots. The organism was taking all its energy out ... Then I was aware of the risk. Many things took place. One of my colleagues stepped into a water pool, and his feet were burned inside his shoes. However, I believed it was my obligation to remain. I felt just like an firefighter. Imagine your home was on fire and the firefighters arrived and left as they believed it was too risky. The firemen discovered it during an annual medical check-up after I'd been there for over a period of time. It was later determined to be normal. I'm not sure the exact date it began to grow. I had surgery to remove half of my thyroid gland. The tumor returned in the last year, and I had the second half removed. I am with (thyroid) hormones at present. Around the time of my operation it was passed by the federal government stating that liquidators must endure for about 4 1/2 years to receive a pension and then retire. If you quit even a day earlier, you wouldn't be eligible for any benefits. ... My condition was basically disabled when I turned 43. I was experiencing similar symptoms to epileptic seizures. The blood pressure in my body was soaring high. It was difficult to be a worker for more than six months in a year. The doctors were unsure of the best way to deal with my. They wanted me to be placed in a psychiatric hospital and label me insane. Then they realized it was due to the radiation."

At the point that the facility was safe enough to conclude the work around 600,000 workers and women had thrown their weight into tackling the aftermath that was created in just several minutes due to the explosions. In the end the surrounding area surrounding Chernobyl did not look the way it did in the days prior to the catastrophe. Sergei said, "I had a chance to visit Pripyat's. It was a spooky experience - a real ghost town. Dogs abandoned (I haven't met cats yet) I'm sure they were eliminated by larger animals) ...) were extremely dangerous in the past. I also had the chance to meet some former station employees in the streets as they scavenged their possessions in the city despite the ban of the government. It was a heartbreaking experience. My cousin was in ChAES for a time as an electrical engineer. his son, who is my nephew, fell sick (they were at home the Saturday... within Pripyat') and up till now, he's not able to recover. When I been to Pripyat in the year 2000, there were no one allowed to reside there. There was barb wire everywhere and MPs were scouting the streets that were empty on BTRs. If I remember accurately, it was about mid-August. However, I could be mistaken. There were many who did not believe that radiation was a real danger. I observed people fishing in rivers contaminated by radiation collecting and eating the contaminated food items like mushrooms, potatoes, etc. The ones who were scared fled the area too quickly. The ones who remained were devastated and extremely upset by the lack of assistance - at any degree. Locals described the incident as the war'. '..."

Timm Suess' picture of Pripyat's "Red Forest"" damaged due to Chernobyl's radiation fallout together with the radioactivity warning symbol

Image of an abandoned house close to Pripyat

In the beginning, Soviet Union would not actually admit how devastating Chernobyl was. Chernobyl disaster was, until the radioactive fallout began moving throughout Europe and increased readings enough in Sweden to let people know that something had occurred. After the catastrophe, Soviet investigations would be followed by international investigations and, although the IAEA concluded that the incident was the result of human error, which was exacerbated due to a lack of expertise and training, Soviet investigations determined the explosions were due to mistakes in design.

However, whatever the reason, the staggering amount of contamination left vast areas of Soviet Union uninhabitable in the days following the catastrophe in addition to the impact it had on animals and the people living in the region. The debate about the extent of damage continues since scientists and doctors remain debating the extent to which people affected and who will be affected by the ailments caused by radioactive fallout. It's been determined that the region in the vicinity of Chernobyl is likely to be affected for at least 100 years.

Chapter 10: The Life-Changing Event

Over 30 years have been passed since the world's most devastating nuclear catastrophe. Chernobyl was once thought to be a major cultural hub but is now an abandoned zone in the north of Ukraine. In the case of Pripyat the town is suffering the same fate as it appears to be an abandoned town. In the future, around 300 years the areas will be empty and polluted, even though there are still people who have chosen to reside in the area.

The word Chernobyl can actually be Ukrainian meaning "mugwort." It is the most common name for an herbaceous species. However, some claim there's a different origin to the term. It's a blend of the terms chornyi as well as byllia. They literally refer to "black stalks" or "black grass." In the 13th century Chernobyl was a city. Chernobyl was among the villages that were crown that belonged to the Grand Duchy of Lithuania. In 1569, the city was made a area of Poland. After Austria, Prussia, and Russia made a series of partitions to break up Poland, Chernobyl became part of Russia's Empire in 1793.

At the end of the eighteenth century Chernobyl transformed into an Hasidic Judaism center. However, in the early period of the 1900s the Jewish residents of the city suffered tremendously during the time that it was the time that the Black Hundreds killed a lot of Jews in Russia. In the conflict that fought between Soviets as well as the Polish people, in 1919 The Polish Army invaded the city of Chernobyl. Then it was the turn of they were attacked by the Red Army invaded the city. The year 1921 was the date Chernobyl was incorporated into the Ukrainian Soviet Socialist Republic (USSR) and the USSR government put in place the nuclear power plant in the year 1977.

On the 26th of April 1986, a sudden surge of power was observed while plant's operators were conducting a system test. As they attempted to trigger the emergency shutdown, it led to an increase in the power surge. This resulted in the explosion of Reactor 4. A second explosion occurred within a couple of seconds as did the igniting of a fire, all that released material in the air which were extremely radioactive. The release of nuclear debris at the facility was 400 times greater compared to the nuclear bomb in

Hiroshima. The theory is that the catastrophe was caused by mechanical and human error.

At the time the disaster struck the population was estimated at 50,000 inside the village of Pripyat. It was not until after around 24 hours when the Soviet government issued an order for the evacuation of the inhabitants. At that point many of the residents were already exposed to different levels in radiation. When they were evacuated, the rationale offered to the residents said that the evacuation was only a temporary circumstance and they weren't required to take their belongings in their luggage. However, the majority of residents were never able to return to their homes. This means that their belongings remain in the same place they were. They are now a reminder of lives that were changed and disrupted in a flash.

Over the course of ten days it released massive amounts of radioactive particles and other substances. In order to address the problem authorities decided to create an enclosed structure known as the sarcophagus, which was designed to contain these substances. The sarcophagus was able to capture around 200 tonnes of trash as well as fuel, which melted straight through the floor before hardening. In May, around 116,000 residents living in the area were asked to move. Then, in July of the same year, another 28 had already been confirmed dead as a result radiation exposure.

Over the following decades, more and more residents were been asked to relocate to areas with less contamination. An exclusion zone was erected that covered 30 kilometers. The majority of relocated residents moved into the town of Slavutych. The city was built following the Chernobyl catastrophe to house those working at the power plant and their families. As of now the city is strictly prohibited from any business or residential activity in the area. The only things that are permitted are monitoring of power plants and installation to study nuclear safety.

At present, there are about three hundred employees in the zone of exclusion, even although they aren't residents. They undergo regular screening to determine if they're exposed. They are also only permitted to work only a few hours every week. They are needed for the facility because the nuclear fuel remains in the remaining three reactors and the fuel requires regular inspection. Some residents, who are mostly older, have refused to leave the zone and others returned in a way that was illegal.

As we mentioned earlier, visitors are now able to take tours of the area. It is possible to schedule scheduled tours in groups or private tours. After the tour, every participant is required to take tests to test for radiation. On the other hand anyone who is concerned over radiation exposure, or poisoning should avoid the tour in any way. While these tours are in place for a long time however, that doesn't mean they're not the risk of.

These 200 tonnes of fuel and nuclear debris that have gotten hard are too radioactive for scientists or researchers to are unable to even get close to them. There are certain elements found in nuclear fuel radioactive that are likely to decay quickly however there are other elements that have extremely long half-lives. Researchers estimate that it could require up to 13 half-lives before living and economic activity could return to any of the affected regions. This means that the entire region that was affected by the catastrophe remains in the same condition for up to 300 years.

If you consider it this event has had such an influence on many people. The Chernobyl accident was an event that changed lives and was not something anyone would have ever. It's now over and over, many are left to consider how this disaster could have been prevented. Of course, thinking about it isn't going to make any much of a difference. The damage is already done. The only thing could be done is find out more about the incident examine what transpired, and then find a way to prevent a catastrophe from occurring again.

Myths regarding the Chernobyl Accident

April 26th 2019, marked the 33rd year anniversary since the tragedy that occurred at Chernobyl nuclear power plant. Researchers, scientists and researchers from all over the world have been trying to stop the negative effects of the most devastating man-made disaster ever recorded. Though more and more facts and details regarding Chernobyl are being revealed in recent several years, there remain some myths floating around that in reality, a large number of people are influenced by. To be able to fully comprehend and appreciate Chernobyl's catastrophe in full you need to distinguish the myths from the facts. There are a few myths surrounding the catastrophe that you should be able to put aside:

1. The Chernobyl accident had devastating health impacts on tens of thousands.

To begin to get you started, here are some figures concerning the health effects directly connected in the event:

* The following day, after the explosion, 134 persons that were at four units were found to be suffering with radiation illness. In addition 28 of them passed away within a few months after the incident while 20 more passed away due to various causes over the following 20 years.

* Over the last 30 years there have been 122 cases of leukemia that have been discovered in workers cleaning up Chernobyl and 37 of those cases could have been caused by radiation exposure.

* Between 1986 and the year 2011 40,000 of 195,000 workers who worked on cleanup in Russia were killed by various reasons. Additionally, the mortality rates were not much higher than those across the nation.

* At the end of 2015 95 of all 993 thyroid cancer cases among teens and children could be caused by radiation exposure.

Apart from that the above, there have not been any other known adverse effects. This is in contrast to the widely-held notions and myths regarding the extent of all radiological impacts caused by the accident to the health of the people. Three decades later, scientists have confirmed the findings.

2. The Chernobyl accident left terrible genetic repercussions.

Nearly 20 years after the devastating Chernobyl incident and Chernobyl's tragic accident, the International Commission on Radiological Protection decided to decrease the chance of a similar event by nearly 90 percent. This decision was based on their belief that they couldn't find any reason not to investigate the possible genetic effects of the incident. It means that all the stories about these effects have been dismissed as bogus and fantasy tales.

3. The Chernobyl catastrophe had a greater impact on the natural world than humans.

The catastrophe caused an enormous radionuclide release into the environment. This is why it is regarded as the most devastating human-caused disaster in history. As time passed the radiation levels remained the same throughout the world but not in the areas where the greatest contamination occurred. The principle of the field of radioecology states that if humans are able to protect themselves adequately and are able to do so, then is the case for the environment, which has numerous redundancies. This means that , if the effect of radiation on humans isn't too extreme, the effects of radiation on the environment will be lesser. The effects of the accident for the surrounding environment evident in the regions around Reactor 4 and which is the Red Forest. The ecosystem is experiencing a complete recuperation. Actually, it's in full bloom!

4. The evacuations were poorly planned and not well-organized.

Although the evacuation of residents of Pripyat was not carried out according to the "ideal" manner however, it was done efficiently and swiftly. Additionally, it's not the case that these residents were exposed to radiation at high levels in the course of their evacuation. In all likelihood, the evacuation was successful. If there was a problem to be pointed out the issue would be that the evacuation took place in a way that was a little late.

5. The government concealed the truth from the public even though they were aware of the seriousness of the situation.

Although there's some truth to this tale however, the situation is far more complex than what it appears. It is certain that the government took the decision to "hide the truth," it may have been due to the fact that they were unable to evaluate the situation quickly. The issue is that Ukraine was not able to establish an effective and independent system to handle such issues. The country didn't have the opportunity to collect information on the radiation levels in the areas that were affected by radioactive fallout. However, if this kind of system was in place at the time, the scenario could have been different.

6. Massive sums of money have been earmarked for aid efforts to help victims of the Chernobyl catastrophe.

In 1992, Russia had spent approximately $3.5 billion following the disaster. The money was used primarily for social welfare. In reality, the amount isn't much since it's approximately $1,000 per victim over 20 years. This amount was adequate to cover the risk. But the reality is that the disaster reduced the growth rate of the nuclear energy sector and other nations.

7. Nuclear power is responsible for the catastrophe since we don't have control over it.

It's crucial to recognize that human error was the main factor in the tragedy. The employees on the site actually violated the rules and guidelines as they completed their safety tests. This is why experts spent years working to improve safety measures and ones that deal with" the "human aspect." This means it's not the blame for nuclear energy. Human error as well as mechanical malfunctions were responsible for the incident. Nuclear power was the cause of the negative effects.

8. The Chernobyl catastrophe was the very first major nuclear disaster involving power.

It is true that the first major accident was recorded in Pennsylvania at Three Mile Island back in 1979. The accident occurred after one of the water reactors that were pressurized suffered a partial meltdown in the core due to human error as well as technical malfunctions.

9. The threshold values for acceptable doses of radiation used by Russia are extremely high.

In contrast the threshold values established by Russia in relation to permissible radiation dosages is among the most stringent anywhere in the world. They quantify dosages of radiation in becquerel (Bq). In the case of milk, the country has the normative dosage of cesium-137. It should not exceed 100 Bq for every Liter. In Norway their accepted dosage is 350 Bq per kilo of food for babies. In this case the milk that has 110 Bq is considered to be radioactive in Russia however in Norway this amount is only a third of the accepted dose!

10. All governments around the globe have retreated from nuclear power due to the disaster that occurred in Chernobyl.

Ten of the top nations in the world generate more than 80 percent of the world's nuclear energy. As of now, Russia seems to be trailing behind the other advanced nations that have already put in place their programs to develop nuclear power. Other emerging countries have expansive programs focusing on the growth and use of nuclear energy. The world today is witnessing a revival in the field of nuclear power as a majority of the world's leading nations have realized that only this kind of energy can address the problems of climate change as well as sustainable development and environmental protection. Even though the Chernobyl disaster was truly terrifying however, it did not stop people from studying and making use of nuclear power. It was instead an important lesson should be learned from to avoid similar events in the near future.

What was the reason the Reactor started to explode?

Today the explosion at the RBMK reactor is the biggest nuclear catastrophe that has ever occurred. Many people question how this event took place. What was the exact cause which caused this reactor go off?

While the exact details of what transpired on the 26th of April 1986 are now long forgotten however, scientists have a rough idea of the way the events transpired. The power plant that was located in northern Ukraine was the site of the four RBMK nuclear power stations. Based on the Nuclear Regulatory Commission in the United States, these reactors located in Chernobyl had power that was high and used graphite for maintaining the chain reaction, while the cores of the reactor were cooled with water.

At the time that the accident occurred at the time of the accident, it was reported that there had been 17 RBMKs operating in the Soviet Union, while there only two in Lithuania. Following the catastrophe has occurred, all remaining reactors located in Chernobyl as well as one RBMK reactor located in Russia and two of those in Lithuania are completely shut down. According to the World Nuclear Association stated that the designs used to build the RBMK reactors was prone to a variety of flaws, and they were a factor in the tragedy. The RBMK nuclear reactors function in the form of "light carbon water reactors." They heat water to generate steam. This is required for the powering of turbines as well as to generate electricity.

The graphite found in nuclear reactors serves as a regulator, helping to maintain the chain reaction, while water functions as cooling agent. This was the biggest issue with designing the RBMK reactors that made it different from other types of power reactors around the world. Based on the Chernobyl catastrophe, many aspects of the design features of the RBMK were unsafe. Particularly the most dangerous elements included those that had a positive coefficient of void as well as designs of rods for control. Following the accident, there were important changes to the design to correct these flaws.

In the evening of the explosion, the team who was working at Reactor 4 was instructed to turn off the power to the reactor quickly in order for that safety check to move through. The purpose of this test was to test the capability that the nuclear reactor could sustain fluid flow in case it be shut down by power. The ideal scenario is by turning the turbines of the reactor ahead of the moment the backup generator is turned on. It wasn't the first time to conduct an safety testing. In the past, tests failed to yield positive results.

The moment the crew realized that cutting off power to the reactor was not a good idea the crew attempted to stop it. But at this point the reactor was in a condition that was very unstable. A peculiarity of the design of the control rods was that, when returned to the reactor produced an enormous power surge. Within the nuclear reactor the control rods could enhance or reduce the rate of nuclear fission for plutonium and uranium by the absorption of neutrons from rogue neutrons.

In the course of nuclear fission rods emit radioactive neurons with the intention of striking other particles of uranium that, in turn, split into lighter elements and start emitting energy. While the rods that controlled the process at this specific power plant constructed of boron carbide, which is neutron-absorbing with graphite tips. Unfortunately, it was graphite that was the substance that resulted in a rise in the fission rate. This particular design flaw was the reason for the massive power surge.

A power surge caused nuclear fuel rods to become hotter and more hot. As this occurred, rods turned the water inside the reactor's core into steam. When the pressure inside the reactor reached a critical level and the reactor was destroyed, it exploded and blew out the cover plate, which was weighing 1,000 tons. This resulted in all the control rods in the reactor to freeze up, even though they were not completely inserted into core. The ferocious production of steam began to be distributed across every part of the. It was fueled by the water flowing into the core due to the fact that an emergency cooling system had broken. This caused the steam explosion as well as the release of fission-related products into the environment.

After a short time had been passed, another explosion shattered fragments of the fuel channels and hot graphite. There are a variety of opinions on the nature of the second explosion. The most likely cause could be the production of hydrogen due to the reaction between zirconium as well as steam. According to reports, when the core of the reactor was exposed to the sun, it released approximately 5 percent of its nuclear material into the air.

While human error wasn't the sole cause of the incident however, it played major part in the incident. The plant's operators began the safety testing in the fall of 2011, they reduced their power consumption of the plant down to 720 MW although they were aware that operating under 700 MW were prohibited. The energy output from the nuclear reactor continued fall to 500 MW until it abruptly dropped to only 30 MW. The engineers began to redouble their efforts to raise the power output to 200 MW or more however, the safety measures were already in place. One reason was that it was required for the reactor to contain at least 15 control rods within the core. But only eight of them were in the core in the course of the experiment.

In the past, analysis and investigations came in the direction of concluding that the root cause of the accident was an explosion of steam. The accepted explanation for the disaster ever since. However, scientists have come to new conclusions. There were many incidents that took place on April 25, which contributed to the tragedy. The owners continued to operate the reactor with lower power and was, as we know was unstable.

The plant's operators attempted to carry out an experiment to create positive feedback loops to create steam and produces energy. Another cause of the catastrophe was the interfering with the automated control system for the reactor. It was this system which inserted the control rods and maintained an unreliable power supply. It is unclear why the emergency shut-down system was activated. The result was the complete insertion of rods that had graphite tips. This caused an increase in the volume of water coolant which caused the massive power surge.

The increased heat and pressure caused tubes which contained fuel to break. Experts believe it was during this time that the first explosion happened and caused the lid to fall off of the reactor. There are those who believe it was more likely the first explosion was a nuclear.

This is due to the fact that when scientists looked at the isotopes they found that they were created through nuclear fission. This could be a sign of an explosion in the nuclear realm.

The examination further revealed an iron plate beneath the core that measured 0.2 km thick was melting because from the blast. This suggests an explosion that was caused by nuclear power not steam. In addition, there were witnesses who claimed to have observed a blue glow or flash just above the structure of the reactor. This could be another sign of a nuclear explosion.

The Truth is Out There to See!

The Chernobyl catastrophe is a great example of just how awful things can turn out when you mix hazardous technology, arrogance of humans, and even politics. In the 1980s, it was the time when Soviet Union set itself apart from the other nations. Each time they shared details, it was typically not anything more than propaganda to impress. That means that what the Soviets were able to share about the incident was usually in error. So, everyone else needed to search for the truth on their own or figure out what was truthful from the statements of they heard from the Soviet Union has said.

Of course, the public is always looking to shed more light on the truth and Chernobyl was no different. In spite of what the Soviets have announced regarding the incident however, the rest of the world knew it was more in the story than they allowed on.

If you attempt to consider what the best scenario for a nuclear accident you could imagine an issue inside the reactor itself. It's easier to imagine an issue that could trigger a chain reaction that can lead to the collapse of the entire system. Who would have thought that such a massive incident could happen during a safety check? This is, however, precisely when and how the accident was triggered.

As we've discussed the various factors that have contributed to the catastrophe itself. One of the most important elements was human mistakes (a many erroneous choices) and the consequences that resulted from of these mistakes. In the safety test the test went horribly wrong. The power surge that was unexpectedly generated caused an explosion and it only went from bad to worse. Then , when carbon monoxide in the reactor merged with air, it ignited the fire to continue burning for nine days. Meanwhile the cloud of radioactive substances was released from the reactorand and then it landed in our atmosphere.

When people think of the effects of radioactivity people start thinking of bizarre things like three-eyed fish as well as other weird things. When you visualize the area around that of the Chernobyl Nuclear Power Plant you may begin to imagine bizarre animals and individuals with physical mutations. Although there's some degree of worry regarding the possibility of mutations arising out radiation exposure, they're not as severe as we may believe. Wildlife might have some developmental anomalies, but the most extreme mutations are unlikely to be passed on to the following generations due to the reason that animals that have been extremely mutated aren't likely to survive long enough to have offspring.

It's true that this incident, though bizarre it had a positive impact on wildlife living in the adjacent areas. Of course, exposure to radiation was not the reason for the advantages experienced by wildlife. What made their lives change in the best way is the reality that many people had fled the area because of radiation. This gave the animals an environment that was free of humans and they began to thrive. There's a positive side effect we can observe from the tragedy.

But it doesn't mean the animals in the area are living blissful and happy lives within the zone of exclusion. There are reports of anomalies in some species, including birds with beaks with strange forms. Researchers have even suggested they believe that the animals (and perhaps even birds without anomalies) might have brains with smaller sizes. There is also the possibility that a significant portion of these animals have acquired an adaptation that results in the production of antioxidants. This can help prevent genetic damage.

It's not only the local wildlife that has changed. The trees that were growing within the exclusion zone exhibit slow growth too. Scientists have also observed an increase in the number of insects, including butterflies, grasshoppers, and bees. They also noted that spider populations are going down too, but this could have positive effects for certain people.

There are animals that don't show physical signs of radiation-induced poisoning but they're radioactive. For instance the boar's bodies in the region are contaminated with high amounts of radiation. One of the most bizarre events that have occurred in this zone is the organisms known as decomposers. They breakdown organic matter. Within the zone, things don't degrade "normally." In example dead trees within the zone which haven't changed or deteriorated even after 20 years from the time the incident occurred. In light of all these enigmas The exclusion zone can be extremely creepy, even though wildlife is flourishing in the area.

The following years after the disaster were extremely difficult as that was when everybody assumed the most dire of outcomes. Two people had already died in the blast and the deaths following these were incredibly tragic. However that it's impossible to determine the true impact of the catastrophe on all those who were exposed to radiation. In 1986, experts believed there would be as many as 40000 deaths due to cancer, but it's a small proportion of the population. That means statistically speaking it's impossible to quantify the rise.

There was a dramatic increase in the incidence of thyroid cancer in children who had been exposed to radiation. There was also the increase of cataracts as well as leukemia among workers at Chernobyl. Apart from the above there was no discernible growth in other forms of cancer among those who had radiation exposure.

In spite of all these changes, anxieties and negative consequences There were still individuals who were unable to go home. Particularly there were people living in the zone of exclusion who did not want to leave and who returned despite warnings. Up to 1,000 people relocated back to their homes during the time of the evacuation. About 100 of them remain there over thirty years later after the incident. The majority of them had roots in the region and others lived in the area together with families a long time.

A fascinating discovery made by workers working in the Steam corridors that are located under Reactor 4 could be "black what they called "lava." After observing the phenomenon, they dubbed this phenomenon the Elephant's Foot because of its appearance. The moniker could be quite charming however it could give a fatal dose of radiation within a few minutes. In the past, being near the elephant's Foot for one hour would be equivalent to having four million radiation x-rays.

A decade after the effects of Elephant's Foot had weakened a little. Even though it was there was still radiation, the effect wasn't quite as powerful as it was in after the tragedy. It's still not safe to be in the vicinity of this object as it can cause radiation-related sickness, or death. The most alarming thing concerning the elephant's Foot is that it's still melting. The dangerous part is that if it melts through the floor and ground then comes into contact with groundwater, there are two dire possibilities--first, it might cause another huge explosion, and second, it might contaminate the water supply. That means they cannot simply leave it to the Elephant's Foot on its own. It must be monitored to make sure it doesn't create more problems.

Could the disaster have been Averted?

April 26th, 2019 marks the 33rd anniversary since the Chernobyl catastrophe. The incident was characterized by massive explosions. It received the highest score for severity by the Internal Atomic Energy Agency. The incident required the evacuation of people living in close areas. As a result of the accident, some people were not able to go back to their home. The disaster left a devastating radiation-related contaminant that will last for the remaining 300 years, or more despite the cleaning efforts currently being undertaken.

All the research and analysis done on Chernobyl The question is how could the catastrophe had been avoided?

Although the catastrophe has already taken place and the consequences are already being felt Many people are left wondering what could be done to have prevented the catastrophe. The fact is that this incident highlights the need for skilled personnel to be working in nuclear power stations. These workers should have the broad understanding and knowledge of radiation science and all the equipment and systems within the facility. There must always be skilled personnel on site every day, so that in the event that something unusual occurs, they will be requested to prevent the situation from getting worse. The incident also signified the beginning of the redoubled efforts to train workers on how to prevent similar catastrophes from happening in the near future.

The incident at Chernobyl was caused primarily by human mistakes. While there were other causes that contributed to the accident, they could have been avoided if the plant's managers made different choices at the start. The incident was the result of an unsafe test that went wrong. Then , the situation got worse because due to the incompetence of personnel who were on site during the incident. However, it wasn't over. After the disaster the Soviets employed secrecy and false information, which clearly caused the situation to get worse. However, the truth is that this tragedy could have been avoided with an effective management system, oversight by regulators and education.

Personnel trained for this type of work and the management of the whole nuclear power station is extremely costly. But, the amount that will be spent on creation and upkeep of preventive measures is comparatively small in comparison to the amount that would be spent on cleaning efforts. In Chernobyl the money that was spent on cleaning efforts could have been sufficient to maintain and train a team of highly qualified and experienced reactor inspectors, communications specialists, nuclear engineers health physicists, risk managers as well as other radiation experts. They can handle sudden events to avoid catastrophic meltdowns and catastrophes like those which occurred at Chernobyl.

With this kind of workers with such a large number of people, a nuclear power plant can be the most safe form of energy for different nations in terms of the impact on the environment and health of the public. However, that wasn't the case with Chernobyl. The workers who ran the safety tests during the accident weren't fully prepared for the event. When things began to unravel, they did not know what they had to do. We know exactly what transpired afterward.

People who have researched and studied the incident have suggested that there could have been possibilities that the blast could have been prevented. One example is if the employees at Chernobyl at the time were experienced and well-trained and experienced, they could have been aware of what to do when an incident was identified. Additionally, remember that the safety test scheduled on April 26, wasn't even the only one. There have been previous tests however, all of them failed. If they had done more thorough analysis of these tests and the reasons these tests were unsuccessful, they could have learned to conduct the latest one with success.

Another way that they could be able to have prevented the tragedy is to know more concerning the engineering of RBMK reactors at the plant. Keep in mind this: the designs of those reactors contained an error that was fatal. While the rods were made of made of boron, which can delay reactors, they were also equipped with graphite tips. These tips speeded up the process and caused a surge in power that resulted in the explosion. However, the graphite-tipped version seemed to be more affordable and that's why it was preferred to rods made entirely of boron.

The accident wouldn't have happened if there not been AZ-15's system defect. Its design RBMK reactor made it possible to produce massive quantities of energy without costing excessively. The Soviets were , according to reports, aware of the flaw in their design, but they chose to not take action about it. This could be the reason why they tried their best to hide it from the world, because they were aware that if the information was released then the truth would become public. In terms of human-made mistakes, Chernobyl could have been prevented if the rods controlling the reactor inside the reactor did not have graphite tips.

Of course, they're only speculations. It's simple to point fingers or crunch numbers and contemplate possibilities. Because no one knows exactly what happened in the plant in the midst of the disaster, we're left with the question of what could've been avoided or not. While there are a few things that could be done to prevent catastrophe, we must think about whether these measures could have prevented the disaster from occurring.

Chapter 11: The Meltdown

The safety inspection that was scheduled to be conducted on the Chernobyl Nuclear Power Plant was thought to be routine, and that's why the plant's director was not compelled to be present. But, the situation changed for the worse when a massive power surge as well as a increase in steam caused explosions that tore Reactor 4 to pieces. The incident is regarded as the most devastating nuclear accident in the history of nuclear power. It killed 30 people directly including 28 firefighters and other workers who were afflicted with extreme radiation poisoning. According to experts, this tragedy is also the primary source of many cancer cases, though they're unable to determine an exact figure. Even to this day the area around it remains so contaminated that it's considered unfit for human use.

In September 1977 in September 1977, in September 1977, the Chernobyl Nuclear Power Plant located in Ukraine was put into operation and began to supply electricity. In February of 1986 an official from the Soviet officials stated that the chances of possibility that a nuclear explosion was one in a million. When this claim was made, Chernobyl's nuclear plant in Chernobyl had four reactors operating and two reactors under construction. On the 26th of April, a sequence of events transpired which led to the destruction of the reactor and the devastating catastrophe. This is a brief overview of the events we will discuss in greater depth in the following section:

* April 25 26th, 1986

* 1:00 am in the morning. The operators of the plant began to reduce the power in Reactor 4 in preparation for the test of safety.

* 22:00 after lunch: plant's operators removed the emergency cooling system in Reactor 4 to ensure that there was no interference.

* 11:10 at night The plant's operators were given instructions to go through in testing for safety.

* April 26 * April 26, 1986

* 12:28 in the early morning The power began to decrease at a rate that was much lower

than it was expected to be and made the reactor unstable.

* 11:30 early in the day: The power began to stabilize, however at a rate that was lower than what was expected. However, the managers of the plant had ordered the safety tests to continue.

* 1:33 early in the morning, plant's operators officially began the test which was the moment when the massive power surge occurred that caused the explosions.

* 1:28 AM The first response team arrived at the scene, not knowing how hazardous this situation could be.

* 2:15 am in the morning The local authorities held an emergency meeting at which they decided to block all vehicles from entering and out of the town of Pripyat.

* 5:00 am in the morning Officials decided to close Reactor 3 completely.

* 6:35 AM The firefighters were able to put out all flames, with the exception of one in the middle that is the core of the reactor.

* April 27 in 1986

* 10:00 am in the morning Helicopters began to dump boron as well as sand, dolomite lead, and clay in the central.

* 2:20 at the end of afternoon, local officials initiated the process of evacuating Pripyat together with nearby towns and villages.

* April 28th 1986: The atmospheric monitors in Sweden discovered radiation levels at high levels and traced their origins back in the USSR.

* April 29th 1986 US authorities were given a glimpse glimpse at the devastation that was caused by Chernobyl explosion by spying satellite images.

* May 1st 1986: Soviet administration decided that it would go through with May Day celebrations in Kiev despite the risk of radiation.

* May 4 1986 * May 4, 1986: The Soviets began pumping liquid nitrogen beneath the reactor, in an effort to reduce the temperature. Other cleaning efforts had also began.

* May 6 1986 There was a massive reduction in radioactive emissions, mainly due to the

fact that the fire inside the core had been extinguished.

* May 8th 1986: The crew had finished the drainage of the underground of radioactive waters beneath the core.

* May 9 of 1986: The workers began to pour concrete beneath Reactor 4 Then, later they enclosed it in an enormous structure constructed of concrete and steel and called it the Sarcophagus.

* May 14 1986. Mikhail Gorbachev, the Soviet leader finally addressed the general public about the catastrophe.

* August 25-29th, 1986 The conference was held by the International Atomic Energy Agency, where scientists claimed that the incident was caused by human error, a lack of culture of safety within the reactor, and the design flaws of the reactor.

The controversy surrounding the Chernobyl disaster isn't resolved until the present day. In fact it was the subject of an HBO mini-series based on the incident that in its own way, described the events of the catastrophe.

In the event of a disaster

Prior to the catastrophe before the disaster, prior to the catastrophe, the Chernobyl Nuclear Power Station was seen as an engineering model and technology by Soviets. It was home to 4 RBMK reactors that could generate enough power enough to supply power for 30 million homes and homes. A RBMK reactor can be described as a kind of nuclear power reactor which is graphite-modified. This reactor was developed by the Soviet Union came up with the concept of this reactor , and went on to construct reactors for nuclear plants. There are some design features of this reactor which contributed to the catastrophe and that is the reason there were calls to decommission the reactors following the accident. Instead of complying with the demands instead, the Soviets designed the reactors. In 2019 there are ten reactors operating.

In 1986 reactors at the nuclear power facility in Chernobyl contained four reactors currently operating while two more were being constructed. Reactor 4 was the latest addition, had 1,600 fuel rods containing uranium-235, which were extremely radioactive. Due to the unstable nature of uranium-235 atoms, they release neutrons continuously. These neutrons are hitting the nuclei of other uranium-235, which can cause them to release neutrons. The whole process is referred to as chain reaction. When chain reactions happen and the result is the emission of enormous quantities of energy and heat. The heat converted the water into steam , which is used to propel the turbines and generate electricity.

To ensure that the chain reaction is under the chain reaction under control so as to avoid the creation of the nuclear bomb the control rods have to be placed within the fuel rods. Control rods contain an element which absorbs neutrons. Reactor 4 contained 211 boron-containing control rods. When these rods were raised, it resulted in an acceleration in the chain reaction. However, when these control rods were depressed and lowered, it caused a slowing in the chain reaction.

For a better understanding of what happened Let's glance at what happened leading to the disaster. In the previous article we gave a brief summary of the things that happened. We will now examine them in more specific detail:

April 25 26th, 1986

At 11:30 a.m. at which time the reactor's operators at Chernobyl began to reduce the power output of Reactor 4 in preparation stage for the safety testing. They scheduled the safety test to occur simultaneously with the reactor's routine maintenance shut down. The goal is to determine whether the turbines in the plant were able to continue spinning to generate energy in order to sustain running the cooling pumps if an electrical failure occurs. In the end, however that test is the one which led to the destruction of the reactor.

At 2:02 p.m. The plant's operators had disabled the emergency cooling system for the core of Reactor 4 in order that it would not interfere with tests for safety. While this didn't directly cause the accident but it did make the impact more serious. At the same time the shutdown and safety test was delayed to meet the power demands of the area. When 11:10 p.m. The plant's operators were instructed to continue the shutdown as well as the safety test. In this same time the plant's operators had switched shifts and the workers who were brought in appeared to be less knowledgeable. The most troubling part was that the employees did not receive the proper training to conduct the safety test.

April 26 April 26, 1986

The shifts were changing, and more than 170 workers walked into the building from Pripyat. It was also the day that they were scheduled to conduct the safety test which was begun 12 hours prior. In simple terms, the safety test was intended to test the capacity of the plant to keep the cooling capacity of Reactor 4 should an electrical failure occurs. As part of the testing, plant's operators were required to shut down the reactor. The night shift workers were on the job the reactor was operating at just 50 percent.

In the early hours of 12.28 a.m., Alexander Akimov who was the foreman on Night shift was involved in an dispute in a heated debate with Anatoly Dyatlov, the Deputy Chief Engineer of Chernobyl. They were fighting over what they should do with the small amount of power Reactor 4 produced. Akimov claimed that, as was it was stated in the manual the power shouldn't be less than

700 MW. However, Dyatlov maintained that dropping to 200 MW wouldn't be hazardous. Akimov was forced to accept the decision because Dyatlov had a higher position. As the power decreased enough to make it unstable in the plant, plant's operators began to remove most of control rods which was, in fact, in violation of safety regulations. But they were unable to boost the power due to the accumulation of xenon within the core.

at 1:09 a.m., Leonid Toptunov The engineer responsible for the control of the reactor was able to stop the reactor's automated shutdown because the level of water was too low. Then , he took out all the rods controlling the reactor (he left only six) which raised the power of the reactor by 77%. But the reactor was beginning to become unstable. That was the time when the most crucial second and minutes began to slip by.

at 1:23:04 a.m. The safety test was officially launched and it was the moment when the largest power outage ever took place. Around 1:23:40 a.m. The readings revealed the temperatures of the nuclear reactor had risen upwards to 4,650 C. It was nearly exactly the same temperature as that of the surface of the sun. One engineer directly over Reactor 4 entered the room for control. He shouted at those who were there, telling them that the caps on those rods for fuel was bouncing upwards and downwards. It was quite an accomplishment since each rod were weighing 350 kilograms. In a state of panic, Akimov pressed the emergency shut down switch, designed to reconnect all of the rods for control. But the rods stopped half way through, and they didn't completely enter.

At 1:43:45 a.m. The reactor was already operating to 120 percent of its maximal capacity, which caused the decomposition of the radioactive fuel. There was a long and low groan that sounding like a human. This was and then an explosion destroyed the shield of 1,000 tons that covered the reactor. The explosion caused the reaction to be exposed to air and, when it came into the contact of oxygen it ignited within the graphite portion that made up the unit. The exposure to oxygen also resulted in the creation of hydrogen gas after the water of the reactor came into contact with the iron inside tube fuel.

The flammability of hydrogen gas can be extremely high and could end up creating another explosion. This time, the blast caused radioactive debris to be hurled into the air as well as onto the top of Reactor 3. Following the explosion the entire plant completely dark because the air was filled by chunks of graphite dust and radiation. The walls and equipment began to collapse as well. This was the time that dozens of fires began to flare up, among which was located on the roof of Reactor 3. Despite all this chaos the nuclear engineer responsible for the experiment kept saying on the need for Reactor 4 remained intact. Then, he passed away because of radiation poisoning.

In the morning, at 1:25.03 a.m. the first call from Chernobyl dispatcher came in and said, "Call everybody, everybody." The Paramilitary Fire Station Number Two received the initial alarm. First responders raced to Reactor 4, and quickly climbed into the rooftop. Because they could see electrical cables being exposed, the rescuers did not make use of water to put out the fires. Instead, they attempted to put out the flames by using sand to cover the cables. The control room at Reactor 4 two trainees known as Viktor

Proskuryakov and Aleksandr Kudryavtsev were waiting to conduct an evaluation of the damage. They arrived in the reactor's hall, and it was here that they discovered the shield was jammed to the shaft that the reactor is located in. Blue and red flames also flared up inside the reactor. Within minutes, the bodies of both men had been darkened by what was referred to as"nuclear tan "nuclear tanning" after they received an extremely lethal dose of radiation.

In the early hours of 128 a.m. firefighters showed up at the site without knowing about the risk of radiation. Due to this, they didn't consider wearing any protective clothes. Around 2:15 a.m. The local authorities called for an emergency meeting. In the meeting, they took the decision to stop any cars from entering and exiting in the city of Pripyat. The police officers who helped with the roadblocks didn't have any idea about the radiation danger and didn't consider wearing the appropriate clothes.

at 2:50 a.m. All of the workers were identified for, except for one man identified as Valery Khodemchuk. At the time of the accident, Khodemchuk was in the main pump room close to the blast. What his coworkers did not know was that the blast had destroyed him. Valentyn Belokon the doctor from Chernobyl who was taking care of the workers who suffered injuries who realized that all were suffering from radiation poisoning. Belokon immediately called Pripyat Hospital to obtain tablets of potassium Iodide. The tablets were designed to stop the build-up of radioactive iodine that was being absorbed into the thyroid glands of patients.

At 5:15 a.m., Soviet officials took the decision to close Reactor 3, which was shut down at 5:00 a.m. The next day they also closed Reactor 1 along with Reactor 2. By 6:35 a.m. the majority of the flames were gone, with the exception of one in the reactor's core which remained burning for several days. At 7:15 a.m., Toptunov and Akimov were able to enter Reactor 4 to pump water into the already damaged reactor. This action cost both their lives. On the 11th of May, Akimov succumbed to his injuries. After three weeks, Toptunov passed away, too. At 8:15 p.m. The residents of Pripyat were all on a bridge the burning graphite at the plant created beautiful flames. While they watched at the scene, a breeze passed over them, carrying the radiation dose of 500-roentgen. This bridge is called"the Bridge of Death since none of them survived.

April 27 in 1986

At 10:10 a.m. helicopters began flying over the reactor , dropping Boron (to take neutrons away) and lead clay, sand, and lead on the reactor to reduce radioactive emission. Though around 1800 helicopters flew by the reactor to accomplish this mission but no of the substances managed to get into the core. Due to the high radiation levels, personnel on the helicopters understood they were on suicide missions, but decided to go ahead with it.

At 22:00 p.m. The evacuations in Pripyat started, though many believed that the move occurred a bit too late. It had been nearly 36 hours since the incident but the residents hadn't been informed of the incident. The announcements were announced via the loudspeaker. The authorities instructed residents to bring enough clothes and food to last for three days, and they couldn't take their animals. Nearly 350,000 people were forced to leave and many did not get the chance to return. All the pets that were left behind in their homes were killed.

The day before, air monitors in Sweden discovered significant levels of radiation in the air, and they tracked the source to Ukraine. The Soviets were forced admit that an incident had took place, but they did not disclose the true nature of the event. They instead claimed the entire situation was "under their control."

The Affected Areas

The nuclear accident that occurred in Chernobyl on the night of 1986, is a tragedy that people won't forget. After you've learned how the tragedy unfolded and what happened, it's time to know more about the affected regions. How wide did the nuclear fallout travel, and which countries suffered the effects?

In the event that the nuclear power station in the northern region of Ukraine was destroyed, the dangers were a concern for the whole continent. The catastrophe that was caused by a safety check had at least 30 people over a span of months. In addition, thousands of others had been affected due to radiation. After the explosion of Reactor 4.4, the reactor's core burned for ten days while emitting radioactive matter to our environment. As per the World Nuclear Association, they estimated that five percent in the material nuclear inside the reactor was released.

This catastrophe resulted in the most massive radioactive release that was not controlled in the history of radioactivity. Most of the consequences of the emission were felt in Ukraine, Belarus, and the western portion of Russia. In actual fact, around one-fifth of the arable land of Belarus is now inaccessible. Unfortunately, when the catastrophe took place the weather was extremely strong also. The winds carried radioactive fallout across Scandinavia and the western portion of Europe. After a couple of days following this incident the other countries like Sweden have also noticed an increase in level of background radiation. In the words of the WHO more than 20000 square kilometers of territory was affected in Europe.

The exact extent of contamination will depend on the amount of rain that fell when the winds carrying contaminants travelled overhead. For example, Chernobyl released a lot of radioactive strontium and plutonium particles. They were able to fall into the ground as far as 100 km from the reactor. The positive side is that there exist isotopes, like radioactive iodine which have very short half-lives. This means they'd have already degraded, being safe from danger. However, areas affected by strontium and cesium will remain a problem over the coming years. For americium-241 as well as plutonium isotopes lengthy half-lives but are not significant to human exposure.

A problem that persists to this day in Scandinavia and the northern regions of Europe is the reindeer herds that consume radioactive plants. This has rendered them unpalatable for humans. Reindeer that are now radioactive roam freely throughout Finland, Norway, Sweden and Russia. Experts have also detected very low amounts of radioactivity in lakes and rivers located in Germany in addition to Sweden. However, the radioactivity found in these lakes and bodies of water isn't substantial enough to be harmful to human beings. The nations comprising Ukraine, Belarus, and Russia remain suffering the most serious impacts of the disaster.

There is currently an exclusion zone of 30 kilometers in the vicinity of Chernobyl. In 2005 researchers from the World Health Organization came up with an estimate of 5 million people who live in the areas that are the most polluted of Chernobyl. Around 100,000 of them lived in areas considered to be to be under "strict supervision." Even though the nuclear disaster occurred in the past 30 years researchers have claimed that the areas that were contaminated aren't suitable for living for the next three millennia. A particular study by Greenpeace found that the accident resulted in environmental harm that is unable to be repaired. In all of history, no been a single event released such large amount of radioisotopes into our the environment.

The Morning After

When Reactor 4 exploded at the nuclear power plant in Chernobyl around thirty years ago released the radiation cloud which contaminated regions of Europe as well as in the Soviet Union before it dissipated. In addition to this cloud, the accident also created a cloud of confusion and misinformation which was mostly resulted from attempts by the Soviets to conceal the disaster. In particular, at the beginning of the disaster of the year, the government of the USSR did not disclose the facts about what had happened to the general public as well as those living in the nearby regions.

Even so it was still a matter of time before the news came out. Actually they Soviet Union had no choice other than to tell the world at large what happened since there were lots of questions about the incident. As pleased as they could be that it was also a shame that the Soviet Union refused to admit that something like this happened within their borders. Because of this, however they could have ruined their own citizens who they did not immediately evacuate after the catastrophe occurred.

The most basic facts about the Chernobyl accident are known. But what happened after the accident? What were they expected to do following the explosions? Even though the government tried to conceal their true intentions, they couldn't stop those who were in the area at the time the incident occurred. Many of the victims were already dead however, there were others who survived and were aware of the full truth. Through the years they have told their experiences to anyone who will be interested, which has increased consciousness of people all over the globe. Online articles, news articles and memoirs, as well as books and other material were written on the disaster because it was so important and affected such an enormous number of people.

For instance, an Harvard historian identified as Serhii Plokhy published a book titled Chernobyl A: History of a Nuclear Catastrophe in which he concentrated specifically on what happened after the disaster as well as the key figures who provided their own personal accounts of what happened following the catastrophe. Consider, for example, Maria Protsenko, one of the founders of Pripyat. Protsenko was in her post at Pripyat for around seven years. The explosion of the reactor had altered her entire world in an second. Even though the explosion had sprayed the town with a lot of toxins however, it wasn't until the next day that authorities decided to evacuate the residents.

The next day, she was able to transition from being a well-known architect to being required to calculate the number of buses they'd require to evacuate all the residents in a safe manner. However, since she was a dependable technocrat, she completed her job and boarded last bus that left to ensure that all the residents made it out first.

The reality is that the next morning following the catastrophe was not a very exciting one, at least not for the world. While the plant's operators as well as firefighters as well as Soviet officials were rushing to bring the crisis under control, the rest of us was living their lives without knowing that something major was taking place in Europe. Even the inhabitants of Pripyat were unaware of what was going on until evacuations began on the afternoon of the 27th of April. However they were told they could return within a few days, and that's why they didn't bring many essential items.

When the news broke the horror that was the incident and the severity it was woken everyone else. And then, everyone wanted to know more details about the incident. This is the reason why there is an ongoing fascination over the tragedy 33 years after it occurred. It is impossible to imagine how the residents of the immediate vicinity were affected, what was going through their heads and what they would do after they heard about the incident.

Real-Life Horror Stories

The fact that the Soviet government believed it was necessary to hide the tragedy shortly after it occurred however, that doesn't mean all of the people living in Chernobyl, Pripyat, and the other affected regions felt the same way. People who survived this catastrophe each have their own story to tell, and do not feel ashamed or hesitant to share their experiences with all of humanity. If you've only had a brief glimpse of the tragedy one of the best methods to find out more and gain an understanding of what transpired is from people who lived through it and who were the most affected by the tragedy.

Lyudmilla Ignatenko

Lyudmilla Ignatenko , the spouse of Vasily Ignatenko, one the firefighters who perished shortly after the catastrophe. Despite the warnings, Ignatenko had decided to be a watchful eye over her husband. Every evening, she would lie in a chair beside husband and attempt to not concentrate on the hurt she was experiencing due to not being able lay down to sleep. A few days ago, a close acquaintance of hers named Tanya Kibenok pleaded with her to go to the graveyard because it was the time of an occasion for her husband's funeral. Vitya Kibenok. She also pleaded with her to go along friend, Volodya Pravik. They both who were close to Lyudmilla's husband.

Following the funeral, as she returned in the hospital she was informed that her husband was dead. The nurse said that according to her the last words spoken by her husband were his own name. And when the nurse informed him that she had left for a short time and was coming back shortly then he smiled and that was it for the day. After hearing this tale, Lyudmilla escorted her husband to the grave however all she remembered of the whole experience was the plastic bag that they put her husband's body in.

While in the mortuary, they inquired Lyudmilla what they'd be dressed Vasily in. While they were able to wear her husband's formal attire but they were unable to put on his cap of service and shoes on as his feet and head were already swelling. This brought back Lyudmilla about her husband's final several days of hospitalization, when the man was choked on his internal organs the flesh was disappearing from his bones. The bandages she used to shield her hands while she meticulously removed his lungs, liver and various organs from his mouth. They continued to come out. It's impossible in her mind, she went through it and, now only has the memories.

On his funeral day she watched as funeral attendants placed the body in an empty plastic bag, and then into an wooden coffin. Then they put the coffin in a plastic bag before putting everything inside the coffin that was made of zinc. This is how all the firefighters who died in the catastrophe were buried - encased with plastic, zinc, wood and cement. There was no way for the families to be allowed to claim the bodies of their beloved loved ones.

Sergei Sobolev

In the past, Sergei Sobolev was the vice-chairman for the committee that was the head of executive of Shield of Chernobyl Association. He was also a member of the committee for funerals. Sergei recounted a particular morning in which the woman was crying and shouting. She wanted them to get all her husband's certificates or medals, as well as benefits so long as they pay the husband back. The woman shouted for quite a while. She also left her belongings to be disposed of, which later will be on display at the Museum. She was married to Colonel Yaroshuk the chemist and dosimetrist, who was dying slowly. He was once as fit in the same way as horses, however, suddenly he was disabled. His wife took care of him and turned him over as required. He had kidney stones that required removal but they couldn't afford the funds for the operation. They made do with what was offered to them. The Colonel was subjected radiation while walking across the zone to identify the areas that were the most contaminated with radiation. He was aware of the job, however he carried it out regardless.

Sergei continues to tell his tales. Who were the people who devoted their time to the roof of the reactor? Over 200 units of the military were tasked to clear the fallout and the units comprised about 340,000 individuals. People who were required to clean up the roof felt the most adverse impacts. Even though they were wearing vests made that were made of lead, radiation was emitted from beneath them, and they were not protected in the area. For their footwear, the men were wearing shoes that were made from synthetic leather, as they worked for 90 minutes on the roof that was radioactive. When they were released with a reward of 100 rubles and an award, a certificate and radioactive exposure. These were in all likelihood the most trustworthy "robots" ever utilized for this task. In addition to working from the rooftop, the workers were forced to sleep in the dirt. To keep warm, they scattered straw onto the ground of their tents. The straw was gathered from the stacks close to reactors. The young men were slowly dying from their actions. However, as Sergei stated, if it wasn't for them, events could have gone differently. These men were from an enthralled culture and were aware that they had to sacrifice their lives to ensure the welfare of the of us.

Nadezhda Petrovna Vygovskaya

Nadezhda Petrovna Vygovskaya was among the people who were evacuated from Pripyat. When they first heard about the evacuation the first thought that came into her mind was "Whose blame was this?" But the more she found out about the situation and the more she began thinking about what she could do about the situation. Nadezhda and those who had been evacuated from the town quickly realized that the evacuation wouldn't only last for a couple of days as they were told, but would last for generations. The realization caused them to take a look back and flip the pages.

As Nadezhda remembers, on the day of the evacuation no one was aware of anything. She took her son to school, while her husband went to the salon for haircut. She was making food when the husband returned home. According to him it was "some sort of fire inside the nuclear power plant" so he advised they take a listen on the radio. At this point, Nadezhda could still see the light of the reactor which was bright and crimson-colored. It was clear to her that this wasn't an normal explosion. As the explosion erupted the people rushed out of their homes to check out the situation. They stood in awe and in awe while they talked, watched breath, and exposed themselves to the threat that was invisible. According to Nadezhda the incident had an odor in the air that caused her throat to itch and caused tears to pour out of her eyes.

At night, she couldn't get a restful night. She could hear her neighbours out and about while they were doing things to entertain themselves. They was taking Citramon tablets to combat an ache that seemed to be brewing and then drifted asleep. When Nadezhda was awake the following morning, she was feeling like something was not right. Though she didn't contemplate it during the time, she could feel that something had changed. Military personnel began arriving at 8:15 a.m. And they had gas masks on. The sight of them brought Nadezhda and the other residents to calm down. They believed they would be safe as the army was coming into town to aid them. In the past, Nadezhda didn't understand that danger was already threatening them and was threatening to kill them.

One thing that sticks out in the Nadezhda memory is the time they'd already gotten on the bus, and everyone was crying. One of the men at the front shouted at his wife calling her stupid, as all she had brought were empty bottles. The wife reacted. She explained that because they were taking the bus, she took her mother the bottles to use. All the other people had huge bags stacked right next to them. everything contained in the sacks was all they had to take when they traveled to Kiev.

Marat Filippovich Kokhanov

At the time, this person was chief engineer at The Institute for Nuclear Energy of the Academy of Sciences in Belarus. According to what Marat recalls, after a month was been passed since the accident, they began receiving materials that were from the 30 kilometer zone in order to conduct tests. They worked for hours, like they were working in a military facility. The pressure was very high since at the time they were the only institution with the necessary equipment and experts across Belarus.

Marat recalls receiving organs and other parts of the insides of unomesticated as well as domesticated animal. They analyzed the milk produced by cows as well as the flesh of different animals, too. After receiving the results from the initial tests, they realized that the flesh could not be considered as meat any more. It was made up of radioactive byproducts. Within the 30-kilometer area there were individuals who cared for the herds of animals during shifts. Shepherds rotated shifts often, and milkmaids were also present according to specific times to milk. But when they examined the dairy, they found that it wasn't more milk. The milk was already radioactive byproduct.

Following these results came out, the dairy in Rogachev began to produce condensed milk, concentrated milk as well as dry powdered milk which the inhabitants had relied on for a considerable period of. These were the same products that were sold in shops. However, when people noticed that the products were made by Rogachev the company, they did not purchase the items. Then, milk products with no labels were spotted on shelves. This was not because they didn't have enough paper that they could print their labels on.

In the year that Marat was first visiting the area when he first visited, he conducted an assessment of the background radiation levels and found that the forest contained levels about six times more than those in roads or fields. However, he was able to detect radiation levels that were high all over the zone. Farmers were digging up their areas, tractors were running and, in a handful communities, Marat and the rest of the staff members at the institute took test of thyroid activity in both adults and children. The results were hundreds of times more than the allowed measurements. One woman within their group, who was a radiologist was hysterical when she observed children playing in the sandboxes. Marat recalls they had checked the breast milk, too and found it radioactive. The majority of the things that were found in the shops in the villages were just radioactive by-products.

In the past, Gorbachev was always on the news trying to calm the masses down. Gorbachev claimed the state had made the necessary changes as well as Marat believed that. He was an engineer for a long time and had a thorough understanding of the principles of physics. He understood that all living things would eventually leave the earth at the very minimum some time. However, they kept taking measurements and watched the news on television. The people of the time were used to believe in the words of their leaders, since they were raised in this manner. They did not realize they were in fact, hearing more truth than what they were taught.

In spite of everything Marat and the other employees understood they remained quiet. They took measurements, but did not reveal the findings to them. They prepared reports, including explanation notes, and carried on in their work without causing any controversy. Why? because they're communists. They did not refuse when told to stay within the zone since back then they had an intense faith. Marat was also a believer and that kept him working for a long time.

Nadezhda Afanasyevna Burakova

Nadezhda had been one of inhabitants of Khoyniki which is a village located in Belarus. Following the incident, when they'd all been educated about it and its effects the effects, she was reminded of a time in which her daughter informed her that in the event that she had a baby with a defect she would be able to cherish the child. This was among her most memorable memories since at the time, her daughter was in the 10th grade. A couple Nadezhda was acquainted with had recently born a boy who was the first son of the handsome and young couple. The baby didn't have ears however, what it did possess was an open mouth. extended beyond the point where his ears were meant to be.

Nadezhda could have left her husband as well as the rest of their family. They considered moving, thought about it, but ultimately made the choice to stay. They were terrified of moving out, like the majority of the others who decided to stay. Nadezhda said that they were not worried about the other people around them. If she was given an apple that they had picked from their gardens she would take the fruit and then consume it. She wouldn't keep this present in her pockets and then throw it away in the absence of anyone else. Every person in the village shared the same experiences and all were destined to the same destiny. To the majority of the world they believed the difference between them and "foreign" as well, or in the worst case, "lepers."

Nadezhda recalls the time when people began using the words "Chernobyl refugee," "Chernobyl youngsters," and "Chernobylites" to describe the people. They were feared by many even though they had no idea about them. While people may not admit to their concerns, Nadezhda knew them as she was a part of the incident. In the initial days following the disaster, Nadezhda brought her daughter to Minsk to find shelter together with her sister. They ended up sleeping in some train stations as her sister would not take them into her home, as she was nursing her infant.

After the accident, bizarre thoughts began to run through her mind. She contemplated doing suicide to make sure they didn't be afflicted, but this was just in the initial days following the incident. People began to think about developing terrifying and inconceivable illnesses. Nadezhda had a degree in medicine and she was able to guess the terrors that other people were imagining in their minds. Everywhere she took her children the people treated them as if they were strangers. She recalls an incident that her daughter attended an early childhood camp in the summer. The other kids in the camp were afraid to be near her. They would tell them things about that they claimed that she "glowed in the darkness" and claimed that she was nothing more than an "Chernobyl rabbit." Her daughter also began to believe in these lies. She would venture out at night to see whether she was, in fact glowing.

The most sad thing concerning Nadezhda and the other survivors is that they're fearful of everything and everybody. They're scared for their own children, for their parents as well as for future generations that don't exist as of yet. People aren't smiling as often. They don't have the same joy like they used to do earlier in their lives. The character of their nation is changing, too. There's a feeling of depress and a constant sense of doom. For those who feel it, Chernobyl has become a symbol or symbol, something that has changed their thinking and the way they conduct their daily lives.

Chapter 12: Research And Studies On Chernobyl

After more than 30 years since the Chernobyl catastrophe Scientists and researchers are still trying to figure out the cause of the disaster, and the ongoing effects that persist today. The disaster resulted in the largest and largest release of radioactive substances into the atmosphere and affected many people, mainly those living in Ukraine, Russia, and Belarus.

A variety of studies have been conducted on the catastrophe, from adverse health effects to how it occurred and much more. The studies have been taken on various topics due to the fact that researchers and scientists have different priorities. Here are a few most popular topics that researchers have delved into in the last few years.

The Changing Wildlife

When Reactor 4 was exploding and released a massive number of radioactive particles in the environment, and many believed that this would result in disastrous effects on local wildlife. But, scientists have proven that this was not the situation. For instance, one study that proved that there was no evidence that suggests that the ecosystem of the lakes within the area was negatively affected because of the catastrophe. In fact, the region of the lake discovered to be the most polluted was also the one with the greatest concentration of invertebrates.

One of the major negative effects the disaster that researchers have discovered was that the mouse species were more in asymmetrical. In addition, both reindeer as well as the roe deer were found to have high cesium levels, making them unfit for human consumption. Fish were also affected by the negative effects, as well. The predatory species in the zone were larger than those that were outside the zone of exclusion.

Forest Fires

From 1992 until 1994, more than 200 forest fires were reported within the zone of exclusion. These fires pose a risk as they pose a threat to those living within the zone, as well as the firefighters that are required to extinguish them. Studies show that the fires have released radioactive particles that pose a danger to the atmosphere, expanding the danger to larger areas.

Radioactive Contamination

Within the exclusion zone all things are in danger of being contaminated by radioactive radiation. The most frequently asked question is how long the effects of radioactivity last? Researchers are currently studying the rate of release of nuclear particles released from Chernobyl and the speed at which the particles are dispersing. Based on the results of these research studies it is likely that Chernobyl won't be a viable area in the near future.

Food

Radioactive isotopes like cesium and strontium been contaminating foods across the European continent following the catastrophe. But research has revealed that there are steps that residents can do to lessen exposureto radiation, like freezing food items to reduce level of cesium, deboning the fish prior to cooking, and many more.

Costs

The whole disaster is expensive. From the destruction caused by it , to cleanup efforts and more, the Chernobyl incident cost the USSR an enormous amount of cash. Particularly, one study found that to achieve the normal exposure level it would cost 10000 euros per person in Belarus and 13,000 euros in Russia while 24,000 euros for Ukraine.

Studies on Chernobyl's Effects of Chernobyl

Anyone who is aware of the Chernobyl catastrophe would know the fact that it caused a massive impact on the environment as well as on the people, particularly those living within the immediate vicinity. In this part, let's review some studies that were done on the impact of the Chernobyl disaster on humans.

1. The Chernobyl Accident An Epidemiological Perspective[1]

The Chernobyl catastrophe occurred over 30-years ago. In the time of the disaster, the explosion released radioactive material into the air, resulting with the exposure of large numbers of people of people across Europe. The studies conducted on those that were exposed have revealed new and important evidence of the connection between radiation and cancer. In particular, these research studies concentrated on the risks of tumors that develop within the thyroid gland because of exposure to iodine isotopes. These studies are important as they aid scientists in understanding radiation's effects and also learn ways to safeguard people from the effects of radiation.

Today, it is established and widely known that children and adolescents exposed to radioactive substances from the catastrophe have seen a significant increase in the incidence of thyroid cancer due to the amount of radiation they were exposed to. People who were at the highest risk of developing the effects of radiation were those who were young at the time of the incident. It was also suggested that a deficiency in iodine stable may increase the risk.

However, the information gathered regarding the risk of developing thyroid cancer in different age groups aren't as conclusive. In addition, some reports have suggested an increase in the death rate or incidence for non-cancer as well as thyroid cancer-related endpoints that aren't cancer-related. Although some studies yielded results that were difficult to interpret because of its limitations in their methodology however, recent studies that focus upon the liquidators at Chernobyl have given researchers the evidence for an increase in chance from various cancers cataracts and cardiovascular disorders following exposure to low doses of radiation.

2. The Mental Health Impacts of the Chernobyl Disaster [22

Since the beginning of time scientists, researchers as well as experts are studying psychological consequences of various disasters which have struck the globe. In terms of the effects to mental wellbeing, most frequent problems that sufferers who are directly affected by these events include stigma PTSD as well as depression and anxiety. other physical signs that cannot be identified medically. The rate of increased mental health-related disorders that are causing morbidity within the first year following incident is about 20%..

Particularly, the kinds of disasters which involve radiation can be extremely destructive since people aren't able to see the radiation, but are terrified of the prospect of being exposed. Being exposed to radiation has long-term negative health effects. After the catastrophe at Chernobyl Numerous studies were conducted on the liquidators as well as the people who resided in the areas that were contaminated. The studies revealed a double increase in anxiety disorders as well as PTSD among those affected. Additionally, their health ratings were significantly less.

In the case of liquidators (the people who were assigned to take care of the mess) The most important danger factor to take into account was the extent the extent of exposure. In the general sample of the general population the primary risk factor was the perception of exposure to radiation levels that could be dangerous. The results are in line with the findings from the survivors and populations of the A-bomb explosion that occurred within Three Mile Island Nuclear Power Plant. Three Mile Island Nuclear Power Plant.

In the case of children, there were some findings drawn from various ecological studies carried out along with local studies that were conducted in Kiev. However, there isn't sufficient direct evidence on how radiation affects people as well as other types of exposures that are teratologic. There is epidemiologic evidence which suggests there is no link between emotional issues, impairments in cognitive performance or academic performance and exposure to radiation, or the stress that comes with the accident.

So the Chernobyl Forum concluded that the most significant public health issue resulting from the incident was the mental health. The Forum based their conclusion on research carried out on adults and liquidators. The issue is, poor mental health could cause death as well as physical morbidity and disabilities. This is why it is imperative to observe the mental health of those after such events.

3. Research Study TRIO: Genetic Impacts on the Kids of Adults Exposed to Radiation from the Chernobyl Accident[33

Information about heritable effects of nuclear accidents are vital to comprehend the effects from exposure to radiation ionizing by parents. There have been very few studies done with sufficient statistical power to determine the impact of the anticipated magnitude, based on studies conducted on animals. Furthermore, the majority of these exposures aren't prolonged or low-dose, like those that are associated with exposure to environmental pollutants and nuclear accidents. In addition, the latest technology of genomics -

- capable of looking at the genome landscape and transcriptome and examining changes in somatics -- haven't been utilized as of yet.

As per reports regions that are located in Belarus and Ukraine that were exposed the radiation that was released by the Chernobyl catastrophe have a higher percentage of mini-satellite mutations among babies born after the accident. A few studies have demonstrated these results however, none of the studies. To expand on these results and to provide new evidence regarding the effects of radiation that are hereditary exposure to parents, scientists have conducted a study on families with one child in which at least one parent were liquidators (thus subject to radioactive radiation) or was one of the people who were evacuated from affected areas. The goal of this research was to find out the heritable and de novo levels of mutation within the genetic variation spectrum. Researchers were specifically looking at the consequences in children, and tried to trace these effects back to the likely parental source.

REB investigators collaborated in Ukrainian's Research Center for Radiation Medicine to collect buccal cells and blood samples, as well as data on the epidemiology of 220 parents who received preconception doses as well as their non-contaminated offspring. The doses that parents received to their in their gonads, from the Chernobyl disaster until the time when they were pregnant with their children were reconstructed for the evacuatees as well as the liquidators. This study employed modern, state-of-the-art genomic technology to develop an analysis and analysis of the genes of the families. The researchers were able to determine whether the exposure to radiation of the parents is associated with genetic changes in the children.

Studies on the Long-Term Ecological Effects

The impact of the Chernobyl catastrophe on ecosystem and the forests that surround the area is a different subject which has been researched by numerous scientists and researchers. After the disaster the entire area was affected which is now known as"the Red Forest. The trees here taken in high levels of radiation and turned into a dark reddish brown color before dying.

Surprisingly, 33 years after the largest nuclear catastrophe in the history of mankind the natural world in the exclusion zone flourishes. The exclusion zone is still eerily quiet, however it is alive with things. While many trees have begun to grow and regaining their strength, scientists found evidence of higher levels of cataracts and albinism as well as lesser amounts of beneficial bacteria in a variety of species of wildlife in the area. However, since human activity has been removed from the process, wildlife has thrived. Let's look at some research studies on the ecological impacts of the disaster:

1. Time-Dependence of Radiocaesium Contamination from Roe Deer: Measurement and Modelling[4[4

Within the peat bog as well as the spruce forest, scientists have studied how 137Cs moved through the earth to plants, roe dwarfs, mushrooms, and eventually to consumers. The concentration of 137Cs in the activity in the roe deer that are found in the spruce forests has begun to decrease slowly. The concentration was superimposed on periodic peak levels during the fall season, and coincides with the season of mushrooms. Scientists have described this decline that is observed to occur over time with the effective time of 3.5 years. This is due to a drop in the 137Cs that are found in soils, which is used to cultivate grasses.

Any additional absorption by the roe deer in 137Cs occurs in fall (mushroom season) is largely dependent on the rainfall during the months of July through September. It also affects the amount of fungi are harvested during the this season of mushrooms. The model developed by the researchers confirmed their belief that the fungi which grow are also able to access a part of the soil's 137Cs concentration that is no longer available for the plants that graze. However the activity level of 137Cs found in the roe deer within the peat bogs is much more. Additionally, it has a more effective half-life that is around 17 years. This is due to the fact that 137Cs bonds irreversibly to the organic matter in peat bogs.

2. In the forest, there are fires that have occurred in the Territory of the Territory that were contaminated as a result Of the Chernobyl Accident The radioactive aerosols are a source of radiation and Exposition of Fire fighters[55

The research was carried out to examine the redistribution and resuspension processes of radionuclides caused by fires in the regions that were contaminated by the Chernobyl catastrophe. In these studies and tests researchers were able to determine the results of the radioactive content of the aerosols that were dispersed, the levels of radioactive aerosols within the atmosphere, as well as the factors and speed of resuspension as well as the flux and speed of deposition in the different stages of the fires as well as at different distances from the fires.

In the active phase of a fire in the active phase, radionuclide concentrations in the air rise by several magnitudes compared to their normal level. The resuspension factor during the active phase was rated as 10-7-. Likewise, it was found that the rate at which resuspension occurred was a magnitude order with the speed of deposition of one. The impact of any additional terrestrial pollution from forest fires can be calculated in the 10-4-10-5 range.

3. Ecological Half-lives of Sr as well as Cs of Terrestrial and aquatic ecosystems[6[6

This study provides a detailed explanation of the behaviour of 137Cs and 90Sr within food products, feeds as well as other environmental media over the long run. Researchers have measured this behavior with the help of the ecological half-life which incorporates all the factors that lead to decreases in the activities of a particular substance such as erosion as well as leaching and fixation.

An extensive number of longer-term concentrations of the 137Cs as well as 90Sr in these media have already been determined and then evaluated again using the application of standard methods for statistical analysis, with the intention of creating sets of information on half-lives within the ecological. Utilizing the example of unaffected waters and soils, the scientists discovered that the concept of an ecological half-life is a matter of debate in particular in the case where the radionuclide examined is found to have a distribution within the study medium that isn't uniform. The researchers also examined whether mixing and transport within the studied medium are relevant at the time they made their observations.

The impact of the Chernobyl disaster on the economy

The Chernobyl accident also caused a major hit to the economy of Ukraine. In actual fact, as years progressed the price of the incident was increasing. Here are the major reasons why costs increased significantly:

1. The damage was substantial directly resulted from the explosion at Reactor 4.

2. The labor and materials used to stop the flow of Reactor 4 were expensive. The initial structure built to protect the destroyed

structure of the reactor was in danger of falling apart, and was in danger of exposing the environment for contamination again. This is why it was the European Bank for Reconstruction and Development (EBRD) along with an array of donors made the decision to construct a replacement that was the NSC that cost EUR2.35 billion.

1. The government was required to build the zone of exclusion with a total area of 30 kilometers.

2. The residents living in Pripyat and the other areas surrounding the power plant were required the need to evacuate and relocated.

3. Health care was offered to patients who were exposed to radiation. The leak in the containment structure exposed thousands of people to radiation levels that were high. Then, around four thousand children been diagnosed with thyroid cancer following the intake of the contaminated milk. In addition, over 600,000 emergency personnel were exposed too. The majority of them died or were suffering from serious health issues.

4. Around 7 million people get cash benefits In Ukraine, Belarus, and Russia.

5. Research and studies are conducted to discover how to make foods that aren't affected by contamination.

6. The radiation levels in the atmosphere must be monitored.

7. Cleaning efforts are conducted to collect and dispose of radioactive waste that is toxic.

8. There were opportunities costs associated with reducing the use of forests and agricultural land.

9. The plant was shut down. energy at the plant. At the time, all the reactors were closed.

10. The termination of nuclear power programme of Belarus caused a loss that amounted around $235 billion.

The accident happened at the worst timing. The moment it was the time that Berlin Wall crashed in 1990 was also the conclusion of Soviet Union. Belarus and Ukraine were members of the USSR However, after the collapse, both countries had to decide whether they wanted the possibility of independence. Prior to the disaster, Ukraine was the "breadbasket" of the Soviets however the Chernobyl disaster ended this position at an end. Following the catastrophe at Chernobyl the business growth in Ukraine was extremely difficult. In the end what businessperson would start a business in a nation that is heavily affected by radiation? Let's examine two reports/studies that concentrate on the economic consequences of the radiation crisis:

1. Economic and social consequences from the Chernobyl Accident[77

In assessing the magnitude of the disaster and the destruction it caused it is clear that the Chernobyl accident, which happened on the 26th of April 1986, was considered to be one of the largest ever recorded by humans using the atomic energy. Actually, when considering this incident from the standpoint of pollution of the biosphere, it's classified as a global catastrophe.

The blast of Reactor 4 resulted in the release of large quantities of radioactive pollutants into the environment. The exclusion zone as well as the other affected regions around 76,100 square kilometers had been exposed C137 at a rate of 1 and 5 Ci/km2, whereas 28,000 square kilometers was exposed to radioactive Isotope C137 in a quantity which was higher than five Ci/km2. The number of people living in the affected areas was estimated to be 4 million. More than 800,000 lived within areas in which the amount of contamination was higher than 5 Ci/km2.

The catastrophe brought about a massive disruption in the lives of those living in the affected regions and also the economic activity of the various regions in the USSR. In the first year following the event it was reported that the Soviets had wiped out more than 140,000 hectares of agricultural land and more than 490,000 hectares of forest. Additionally, a large number of industrial and agricultural companies were forced to cease operations.

In the summer and spring season of 1986, a large number of residents in the zone of danger were forced to evacuate their homes as soon as they could. In addition to the evacuations, different types of tasks were performed to ensure that the reservoirs safe from radioactive contamination. Additionally, they constructed special traps for hydraulics as well as structures to prevent the silt that is radioactive from moving. The cleaning and cleanup work which were carried out to take care of the effects of the accident took place over three times.

Initial Period (April 1986 until May 1986)

In this time, they came up with the first estimates of the size of the disaster , along with the radiation pollution situation. They had to act to stop the release of radioactive material or chain reactions that could occur from Reactor 4. They also needed to determine the areas which were exposed to radioactive substances and evacuate residents within the 30-kilometer area. In this time, the most significant dangers were posed by exposure to radiation from the outside and also from radiation exposure inside mostly from breathing in or eating radioactive particles that are within the atmosphere.

The Second Time (Summer of 1986 through 1987)

In this time they had already mapped out all the areas that were affected by contamination. They constructed the sarcophagus and decontaminated the area of work and then restarted the operation for Reactor 1, Reactor 2 as well as Reactor 3. They also created strategies to ensure that waters were shielded from radioactivity. They also cleaned the settlements. Research was conducted and agricultural lands were given special treatment. In the second time the most significant risks of radioactive contamination were cesium-137, cesium-134and cerium-141 and cerium-144. the ruthenium-106.

The Third Period (From 1988 until Present)

In the course of this time the NSC had stabilized the situation in the 30-kilometer zone as and the rest of the areas. They had set up an appropriate system of monitoring and dosimetric work, conducted tasks for the NSC as well as removed the settlements of contamination, relocated the inhabitants to avoid contamination, implemented steps to minimize the risk of contamination to crops, and reorganized agriculture activities. They also began gathering information on the accident to formulate an over-the-long term plan for managing the consequences of the catastrophe. Today, the main dangers of radioactive contamination come from cesium-137 and strontium-90 with longer half-lives.

Despite all the efforts to mitigate and minimize the repercussions of the Chernobyl disaster , and in spite of the massive amount of technical, material and financial resources dedicated to these efforts, there's currently no reliable system to guarantee the safety of every person who are affected.

2. What was the economic impact for Chernobyl? Chernobyl Accident? [8]

The nuclear disaster that occurred at Chernobyl as well as the policies by the government implemented to address the consequences, entailed massive cost for both the Soviet Union and its successor nations, Ukraine, Russia, and Belarus. Although these countries suffered the worst consequences of the disaster, other nations also suffered economic losses.

It's only possible to determine the cost of the accident by generating an estimate. This is possible considering the non-market circumstances that were in place at the time the incident took place and the fluctuating exchange rate and high rates of inflation during the period of transition that followed when the Soviet Union ended in 1991. Naturally, the magnitude of the impact of the catastrophe is evident. Indeed, various estimates by the government in the 1990s have put the total cost as thousands of millions of dollars. The funds were utilized to fulfill both direct and indirect goals, and include the following:

Direct expenditures:

* the damages resulting from the accident in and of itself

* actions taken to close Reactor 4 off and reduce the consequences of the accident in the zone of exclusion

* relocation and evacuation of residents, as well as building new housing and infrastructure that can be able to accommodate them all

* Social and health protection provided to the people

* Research on the health of our environment and methods of producing healthy, safe food

* Monitoring radiation levels of the atmosphere and the surrounding environment

* the enhancement of radioecological quality of the settlements , as well as the effort to dispose of radioactive waste

Indirect expenditures:

* the chance cost of clearing land and forests, in addition to the closure of agricultural and industrial facilities

Additional energy expenses due to the power failure in the power station as well as the deactivation from the power nuclear programme of Belarus

The budgets of the national government had borne the most burdens during the course of dealing with the consequences of the Chernobyl catastrophe. For Ukraine alone, each year, as much as 7 percent of the government's expenditures remain devoted to the benefits and programs that are a result of the Chernobyl disaster. As of 1991, the state of Belarus invested more than 22 percent of their budget, but this number was gradually decreasing to just 6 percent by 2002. The total amount that was spent by Belarus from 1991 until 2003 was in excess of $13 billion.

This massive expenditure created an economic burden that was not sustainable, particularly specifically for Ukraine in particular Belarus. Even though the capital-intensive expenses for resettlement programmes have been halted or canceled however, massive sums of money remain distributed in social benefits for millions of people across Ukraine, Belarus, and Russia. With resources shrinking governments must begin cutting back on the Chernobyl-related expenses to offer better targeted and specific assistance, assisting those who are most at danger of being deprived of their socio-economics or health threats.

Conclusion

On the 26th of April 1986, the most devastating nuclear catastrophe occurred within the Chernobyl Nuclear Power Plant located in Ukraine. The site of the power plant was close to the border between Belarus and Ukraine in what was once the Soviet Union. There were four operating reactors which produced 1000 MW of power per and were contemplating the construction of two additional. The fire was caused by the reactors' design flaws and human mistake. The day before the incident the plant's operators were shutting down the power system for Reactor 4 (the that which controlled the power) together with the emergency safety system of the plant in order to conduct the safety test. At the moment, Reactor 4 was running at just 7.7%. In the absence of the plant's owners the defect within the structure of Reactor 4 meant when it was running at lower power levels, the reactor also released more neutrons. The situation got worse when plant managers took out the graphite rods, which were responsible for regulating the release of neutrons.

The choice to insert all the rods at once resulted in the reactor spiraling beyond control. The explosion was so powerful that it destroyed the lid of the reactor, constructed from steel and concrete. The result was the release of huge quantities of radioactive substances and other substances into our environment. Despite the seriousness of the situation however, it was the Soviet Union had decided to keep this catastrophe from the the world and even their own citizens. The radiation however had been the cause of warnings across Sweden and the Swedish authorities traced the source of the elevated radiation levels back to Soviet Union. When the Swedish government Sweden demanded the Soviets regarding this, they were left with only one option: acknowledge that an accident took place within Chernobyl. Chernobyl plant, however at the time, they did not reveal all the specifics.

Two people at least instantly died in the blast and a number of others died because of severe radiation poisoning in the months following the incident. Soon after massive evacuations took place in Pripyat as well as in the immediate vicinity of the plant. In the years following the accident the livestock was affected, and cases of cancer in children grew significantly also. After a few years it was discovered that the Soviets created a containment structure known as the sarcophagus in the destroyed reactor. It was constructed from concrete and steel. Since the structure was constructed quickly it was later found to be dangerous. Radioactive contamination had spread throughout the continent of Europe and affected Italy, France, and other countries. The outbreak also led to protests across the globe about the inadequate security standards, as well as the reactor's inadequate design.

It is believed that an incident that severe is highly unlikely in a nuclear plant located in Western countries , since they need periodic safety checks and protocols. However, all countries are at risk when it comes to the force of nature as we've seen through the tragedy at Fukushima. As the tsunami hit after which an earthquake occurred which resulted in the cooling generators to become removed from the cooling generators. Although they had stopped the reactors the event caused three nuclear meltdowns along with a few radiation leaks.

One of the greatest results that could be gained from these situations is that people can learn from the mistakes. Nowadays the nuclear power industry is more secure since these events occurred. The other reactors of Chernobyl Nuclear Power Plant Chernobyl Nuclear Power Station are not operating, and plans to construct two more reactors are long gone. It is now just a place that is working to dismantle the original structure of containment (the Sarcophagus) in an effort to limit the radiation danger that is still a threat to the other people of the world.

There is no plan to construct nuclear reactors similar as those used in Chernobyl. Chernobyl plant. Additionally, careful considerations are given to site safety when trying select the best locations for new reactors. That means that locations which pose a significant danger to natural catastrophes aren't permitted. The most significant breakthrough is in the new technology used in design. There are currently "pebble bed" designs made of "pebbles" of fuel that are as large than tennis balls. The design of these pebbles permit them to work at extremely high temperatures and without problems. These pebbles are also cooled by gas rather than water-cooled, making them more efficient as gas isn't able to absorb neutrons. According to the design team, they are more secure in every way.

Of course, production of energy by any method is risky in one degree or in another. For instance, hundreds have already been killed due to explosions from oil drilling rigs. In addition, hundreds suffer each year when the dam that is hydroelectric collapses, then floods villages. These disasters cause environmental damage also. To move forward we shouldn't quit the concept of energy consumption and revert back to technology from that of the Stone Age. Scientists are instead working to make the process of producing energy as safe as they can and also learn about what they shouldn't do after incidents and disasters like those that occurred at Chernobyl. When you compare nuclear energy with other sources of energy it's far more secure and less polluting that's why we shouldn't just stop using it. There are a few essential lessons to be taken away from the Chernobyl accident (and other similar incidents) and include:

1. Always be honest.

Following the devastation of Reactor 4 in Chernobyl After the Chernobyl nuclear disaster, Soviet authorities attempted to conceal the incident. They failed to inform people or other nations about the risk posed by the incident for several days, which put millions of lives in danger. In reality, they just taken the decision to remove the residents of Pripyat the closest town to the power station, the following day. This meant that all residents had experienced radiation for a long time, and they did not even realize that there was a hidden threat threatening their lives.

There is the need for full transparency, particularly in the case of nuclear incidents. If something goes wrong and those in charge are trying to hide the truth from the public and it ends up in tragedy.

2. Get out as soon as you can at the earliest time.

At first the Soviets were denial about the magnitude of the disaster. They were forced to wait for a while before making the decision to evacuate the residents of Pripyat as well as the surrounding regions to the safety of. It was the best choice. If the Soviets were to have evacuated the inhabitants earlier, this could have saved the lives of many.

3. Be aware of the radiation levels in your food items closely.

The scientific group from the United Nations released a report on the effect of the atomic radiation. They found that the rise in Thyroid cancer patients was the main and sole medical legacy left behind by Chernobyl. These cancers were triggered due to the fact that the Soviet government allowed for the manufacturing of milk polluted, and the milk was consumed by children. This is why many children were exposed to large amounts of radiation into the thyroid glands of their children.

4. Always follow safety guidelines and regulations.

According to various reports from around the world, the reason of the explosions that occurred at Chernobyl was the flawed construction of RBMK reactors. Other reasons were the facility's unsafe culture, as well as the poor decisions taken by the workers who conducted the safety tests at the time of the accident. The incident is a clear illustration of the importance of observing safety guidelines and regulations in nuclear power stations. Inspections for safety are crucial to allow experts to confirm whether the plant is operating in a safe manner or whether it is necessary to make changes.

5. Prepare for the future.

As per experts, the disaster at Chernobyl has also established the need to establish and maintain an efficient emergency response system for each region. This kind of system is vital when man-made disasters happen.

The Chernobyl catastrophe was an incredibly tragic and terrifying event. Many lives were lost and lives were altered and the world was awakened to real danger that nuclear power poses. However, this doesn't mean nuclear power shouldn't be considered a viable option. So long as we take our lessons from our mistakes and develop plans to safeguard in the near future, then we shouldn't need to be worried about having another catastrophe that is as disastrous like this one.

www.ingramcontent.com/pod-product-compliance
Lightning Source LLC
Chambersburg PA
CBHW050401120526
44590CB00015B/1779